Consumer Perceptions of Tap Water, Bottled Water, and Filtration Devices

The mission of the Awwa Research Foundation (AwwaRF) is to advance the science of water to improve the quality of life. Funded primarily through annual subscription payments from over 1,000 utilities, consulting firms, and manufacturers in North America and abroad, AwwaRF sponsors research on all aspects of drinking water, including supply and resources, treatment, monitoring and analysis, distribution, management, and health effects.

From its headquarters in Denver, Colorado, the AwwaRF staff directs and supports the efforts of over 500 volunteers, who are the heart of the research program. These volunteers, serving on various boards and committees, use their expertise to select and monitor research studies to benefit the entire drinking water community.

Research findings are disseminated through a number of technology transfer activities, including research reports, conferences, videotape summaries, and periodicals.

Consumer Perceptions of Tap Water, Bottled Water, and Filtration Devices

Prepared by:
Erin D. Mackey, **Jason Davis**, **Lina Boulos**,
Jess C. Brown, and **Gil F. Crozes**
Carollo Engineers
12592 West Explorer Drive, Suite 200
Boise, ID 83713

Sponsored by:
Awwa Research Foundation
6666 West Quincy Avenue
Denver, CO 80235-3098

Published by the

Awwa Research Foundation
Advancing the Science of Water®

IWA Publishing

DISCLAIMER

This study was funded by the Awwa Research Foundation (AwwaRF). AwwaRF assumes no responsibility for the content of the research study reported in this publication or for the opinions or statements of fact expressed in the report. The mention of trade names for commercial products does not represent or imply the approval or endorsement of AwwaRF. This report is presented solely for informational purposes.

Published by IWA Publishing, Alliance House, 12 Caxton Street, London SW1H 0QS, UK

Telephone: +44 (0) 20 7654 5500; Fax: +44 (0) 20 7654 5555; Email: publications@iwap.co.uk;
Web: **www.iwapublishing.com**

AwwaRF report number 90944F. Originally published by AwwaRF for its subscribers in 2003.
IWA Publishing version published 2004.

ISBN 1 84339 853 2

CONTENTS

TABLES

FIGURES

FOREWORD

The Awwa Research Foundation is a nonprofit corporation that is dedicated to the implementation of a research effort to help utilities respond to regulatory requirements and traditional high-priority concerns of the industry. The research agenda is developed through a process of consultation with subscribers and drinking water professionals. Under the umbrella of a Strategic Research Plan, the Research Advisory Council prioritizes the suggested projects based upon current and future needs, applicability, and past work; the recommendations are forwarded to the Board of Trustees for final selection. The foundation also sponsors research projects through the unsolicited proposal process; the Collaborative Research, Research Applications, and Tailored Collaboration programs; and various joint research efforts with organizations such as the U.S. Environmental Protection Agency, the U.S. Bureau of Reclamation, and the Association of California Water Agencies.

This publication is a result of one of these sponsored studies, and it is hoped that its findings will be applied in communities throughout the world. The following report serves not only as a means of communicating the results of the water industry's centralized research program but also as a tool to enlist the further support of the nonmember utilities and individuals.

Projects are managed closely from their inception to the final report by the foundation's staff and large cadre of volunteers who willingly contribute their time and expertise. The foundation serves a planning and management function and awards contracts to other institutions such as water utilities, universities, and engineering firms. The funding for this research effort comes primarily from the Subscription Program, through which water utilities subscribe to the research program and make an annual payment proportionate to the volume of water they deliver and consultants and manufacturers subscribe based on their annual billings. The program offers a cost-effective and fair method for funding research in the public interest.

A broad spectrum of water supply issues is addressed by the foundation's research agenda: resources, treatment and operations, distribution and storage, water quality and analysis, toxicology, economics, and management. The ultimate purpose of the coordinated effort is to assist water suppliers to provide the highest possible quality of water economically and reliably. The true benefits are realized when the results are implemented at the utility level. The foundation's trustees are pleased to offer this publication as a contribution toward that end.

Edmund G. Archuleta, P.E.
Chair, Board of Trustees
Awwa Research Foundation

James E. Manwaring, P.E.
Executive Director
Awwa Research Foundation

ACKNOWLEDGMENTS

Carollo Engineers gratefully acknowledges that the Awwa Research Foundation is the joint owner of the technical information upon which this presentation is based. Carollo Engineers thanks the Foundation for their financial, technical, and administrative assistance in funding and managing the project through which this information was discovered. The comments and views detailed herein may not necessarily reflect the views of the Awwa Research Foundation, its officers, directors, affiliates, or agents.

Clearwater Research Inc. (Boise, Idaho), led by Study Director John Hetherington, Ph.D., performed all the market research work for this project. This includes collaboration on survey structure and content, all survey coding, execution and data analysis, and guidance in survey interpretation by Carollo Engineers personnel throughout the project.

The authors gratefully acknowledge Dr. Joseph Coates of Coates Futurists, Inc. for development of the future forecasting work for this project.

The authors of this report are indebted to the following water utilities and individuals for their support, cooperation and participation in this project:

- Aquarion Water Company (formerly BHC Co.), Bridgeport, Connecticut - John Herlihy
- City of Tempe Public Works Dept., Tempe, Arizona - Tom Gallier, David McNeil and Tom Ankeny
- City of Midland, Michigan Water Utility - Vic Chichester
- Colorado Springs Water Utility, Colorado Springs, Colorado - Curtis Mitchell
- Corpus Christi Water Utility, Corpus Christi, Texas - Eduardo Garaña
- Mount Pleasant Waterworks - Clay Duffie
- Orange County Water Utility, Orlando, FL - Jacqueline Torbert and Valerie King
- Philadelphia Water Department, Philadelphia, Pennsylvania - Gary Burlingame
- South Central Connecticut Regional Water Authority, Connecticut - Cheng-nan (Mike) Weng
- Southern California Water Company, San Dimas, California - Denise Kruger, Bill Gedney and John Brady
- Sydney Water Corporation - Naomi Roseth
- United Water Idaho, Boise, Idaho - Camille Brown, Scott Rhead and William Carr
- Village of Wilmette Water Utility, Wilmette, Illinois - Ray Ames, Jr. and Paul Christenson
- Water District N° 1 of Johnson County, Mission, Kansas - Byron N. Johnson, Eileen Koutelas and Charles Weber

The authors also wish to acknowledge the input and guidance provided by AwwaRF Project Managers Kim Linton and Linda Reekie, and the Project Advisory Committee (PAC)—including Liesel Adam, Lehigh County Authority, Allentown, Pennsylvania, James Epp, Kansas City Public Utilities Board, Susan Hall, Kansas City Water Services Department, Kansas City, Missouri, and Elaine Tatham, President, ETC Institute, Olathe, Kansas.

EXECUTIVE SUMMARY

BACKGROUND – TAP WATER ALTERNATIVES

Recently, much attention has been given to the consumption and use of tap water alternatives (TWAs) such as bottled water and point-of-use/point-of-entry (POU/POE) devices. Bottled water can originate from tap water, groundwater, or natural spring water and might be filtered, distilled, deionized, carbonated, or mineralized. POU/POE devices, which are used to further treat tap water, generally rest on a countertop or are installed in-line underneath a sink.

The water market has changed dramatically in recent years. Significant growth in public/private partnerships (20–23% in the last 10 years [Dysard 1999]), intense marketing of bottled water and POU/POE devices, and the recent increase in media attention to drinking water issues have contributed to a shift in public attitudes toward tap water. Data collected in recent years reveal some interesting statistics about the emerging field of TWAs and public perceptions.

- 86% of Americans are concerned about their tap water (WQA 2001)
- There is a 10% annual increase in the production of bottled water (Bottled Water Web 2000)
- 110 new bottled water companies emerged in 2000 (Bachmann 2000)
- 2 million POU/POEs are purchased annually in the U.S. and Canada (Doulton Water Filter 2001)
- Annual bottled water sales are $4 billion in the U.S. & $36 Billion worldwide (Suffet 2001, Seattle Times 2000)
- 41% of Americans use POU/POEs, bottled water or both (WQA 2001)
- For a given geographic location in the U.S., the percentage of people who drink bottled water can be as high as 75%. (Watertech 2000, Suffet 2001)

What is of particular interest about these statistics is the fact that people receive safe tap water every day. In fact, only about 5% of the 55,000 water treatment systems in the U.S. ever exceed any USEPA maximum contaminant level (MCL) standard (EPA 2001b). Furthermore, 75% of all MCL violations are a result of monitoring/reporting practices and are not necessarily related to actual drinking water quality.

OBJECTIVES

Though average American tap water quality is very good, it is evident that an increasing fraction of the national population is turning to tap water alternatives as their primary drinking water source. A continuation of this trend could have serious consequences for drinking water utilities across the nation. Only by identifying and understanding the various driving forces behind this shift in water consumption trends can utilities begin to develop a strategy to compete effectively in the long-term drinking water market. Therefore, the objectives of this project were as follows:

- To measure consumer satisfaction with tap water quality.
- To investigate demographic trends in consumer satisfaction and consumption of tap water alternatives.
- To identify the factors that cause consumers to purchase and use bottled water and POU/POEs.
- To investigate whether these factors are real or falsely perceived.
- To develop a list of recommendations for water utilities that can improve consumer satisfaction and help bridge gaps between perception and reality.

RESEARCH APPROACH

The research approach focused on identifying reasons underlying individual decisions to use alternative sources to tap water. The general approach was to identify customer attitudes and perceptions of tap, bottled, and filtered water in several geographically diverse markets. The specified approach was as follows:

1. Gather physical data on each of twelve target markets across the U.S.. This approach provided a physical data set to which perceptions survey data could later be associated. Utilities varied by size, type of ownership and geographic locations.

2. Conduct a telephone survey to investigate public perceptions of bottled water, POU/POE devices, and tap water in the target markets. The survey covered pertinent issues with respect to water consumption habits and survey analyses were statistically rigorous. Approximately two hundred persons (100 tap water drinkers and 100 bottled or filtered water drinkers) were surveyed in each of the twelve target markets. Quota levels were managed in order to attain 100 respondents in each category.

3. Summarize and compare the survey results between markets and between utility management and consumers. The survey results were compared between markets (using geographical and physical data to define differences between populations) and between user groups (tap water and filtered/bottled water drinkers). These results were also compared to survey results from a utility manager's survey. In this survey, utility personnel, primarily management, were asked to give their perception of consumer views and attitudes.

4. Formulate recommendations for water utilities. Survey results and analyses were used to develop strategies to improve public relations, CCRs, and consumer education programs.

RESULTS

National Survey Results

The aggregate results of all 12 markets represent a snap shot of the national consumer market. Regarding overall tap water quality, tap water (TW) drinkers were 18% and 39% more satisfied than bottled water (BW) or filtered water (FW) drinkers, respectively. Out of six areas defining tap water quality (overall quality, taste, appearance, safety, smell, healthiness), safety retained the highest level of satisfaction by both TW and TWA drinkers (92% and 74%, respectively. The area with the fewest percentage of satisfied consumers was taste (82%-TW, 52%-TWA drinkers).

Survey respondents also described the taste of their household tap water. In each market "offensive" flavors (musty/earthy, metallic, and chlorinous) were reported 3 times more often by TWA drinkers than by TW drinkers. Furthermore, BW drinkers described their tap water as "offensive" more often then FW drinkers. Together, these data suggest that perception of taste may be a major driver for choice of drinking water source, particularly when bottled water is chosen.

TWA drinkers were asked to cite the most important reason for using a tap water alternative. Safety/healthiness was the primary reason for the majority of TWA drinkers (68% and 56% for FW and BW drinkers respectively), while 26% and 30% (of FW and BW drinkers) respectively cited taste. FW drinkers appear more concerned with safety and health of their tap water than do bottled water drinkers.

COMPARISON BETWEEN TARGET MARKETS

Comparison of Bottled and Filtered Water Consumption and Utility Characteristics

Consumption of tap water alternatives was highest in the "West Coast," "Southwest," and "Southeast 1" markets reaching in excess of 80%. The only other comparable survey conducted in the U.S., which examined similar market areas, determined that 40-50% of Americans used tap water alternatives (AwwaRF 1993). Not only does this represent a sizeable increase in less than 10 years, but the drinking water market is now at a point where Americans in some geographic areas consume relatively little tap water.

Geographic variation in tap water alternative consumption statistics can range as much as 60%. While the "West Coast" exhibited a high rate of use at 80%, only 20% of the people in the "Midwest 3" market consumed tap water alternatives. Factors that influence this variation may include local water quality, utility characteristics, and demographics. While other factors such as media influence and product availability for bottled/filtered water also play a role in determining consumption patterns, these factors can be extremely difficult to quantify in a meaningful manner.

Effect of Demographic Differences on Bottled and Filtered Water Consumption

Each market was independently studied to assess whether age groups, gender, education levels, or number of children in the household significantly affected consumer perception of overall tap water quality. Following the demographic analysis, an examination of utility characteristics (e.g., size, ownership, water source, and finished water quality) that could affect consumer perception of the use of tap water alternatives was conducted.

The most apparent trend was found among young consumers (29 and under). This subgroup had decreased levels of satisfaction for tap water safety, healthiness and overall quality. As age increased, so did the level of satisfaction. However, it should be noted that the average satisfaction levels for all age groups were always above neutral. The presence of children in the household did produce some differences in tap water satisfaction levels, but there were no observed general trends. Gender and educational differences were not significant.

Age was a significant factor affecting the choice to drink bottled water while at home in ten of the 12 markets. Respondents 44 years and younger were more likely to drink bottled water at home, whereas older respondents (65 + years) were less likely to drink bottled water at home. The presence of children in the household also affected the choice to drink bottled water while at home. Gender and level of education were not significant factors. FW drinkers showed no consistent usage pattern for any of the demographic factors examined (age, gender, education, or presence of children in a household).

EFFECT OF UTILITY CHARACTERISTICS

Utility ownership, size, source water, and finished water quality all affected consumer perceptions of tap water quality and utility service. Relative to customers of private water providers, customers of public water providers were more confident in the safety of the tap water they receive at home and were more likely to believe that it meets all state and federal regulations. Customers of moderate (5-20 mgd) to large water providers (50-100 mgd) were less satisfied with overall tap water quality, safety, and healthiness, when compared to customers of small water providers (0-5 mgd). Customers of water providers using groundwater as source water were less satisfied with overall tap water quality, safety, and healthiness, as compared to customers of water providers using surface water as source water. Consequently customers of water providers treating surface water were more likely to drink household tap water, while customers of water providers treating groundwater were more likely to use filtration devices or to drink bottled water while at home. A regression analysis showed that 20% of the variation in the use of tap water alternatives was attributable to the use of groundwater as a drinking water source.

Water quality parameters expected to play a role in consumer drinking patterns and satisfaction levels included finished water temperature, hardness, and total dissolved solids (TDS). None of these parameters showed any correlation with the usage rates of tap water alternatives, suggesting that by themselves these parameters are not significant enough to cause people to change to bottled/filtered water. However, TDS and hardness were correlated with water quality satisfaction. Whether examining consumer satisfaction with overall quality, taste, safety, healthiness, smell or appearance, 34-64% of the variation (i.e., R^2-value) in peoples' satisfaction levels was associated with the level of TDS or hardness in their tap water.

UTILITY MANAGERS' SURVEYS

A survey was distributed to water utilities nationwide, covering twenty-two states located in various regions throughout the country. Approximately 130 contacts were made. A total of 117 responses were received from agency representatives associated with 57 utilities. A member of upper management was requested to fill out the survey (e.g., plant manager or department head). Some utilities also had "support personnel" (e.g., operators, engineers and customer service) respond. These persons comprise about 1/3 of the total responses.

The survey was designed to assess utility perception of users' attitudes toward utilities and to complement the data collected in the consumer survey. By comparing consumers' and managers' responses for a given question, a quantitative analysis was done, which showed areas where a utility could improve.

A comparison between responses given by utility managers to consumers revealed that communication between utilities and customers was not optimal. An example of this point is illustrated by the fact that the majority of TWA drinkers were motivated to use tap water alternatives due to safety concerns, while utility managers typically cited aesthetics (taste, smell, appearance) as the primary motivator. However, communication between managers and consumers was not entirely lacking. Three of the 12 market area managers surveyed believed that consumers did remember receiving a CCR (refered to as a "Water Quality Report" in the customer survey). Compared to the 36% from the consumer survey, it would appear managers have a reasonable awareness if not a very good guess regarding this issue of CCRs. Therefore, while communication may be effective here, results show a definitive need; more effective outreach methods are needed. In order to be competitive in a market that now includes private enterprise, it will be critical for utilities to implement new public outreach methods to reach consumers who failed to acknowledge receipt of their CCR.

RECOMMENDATIONS FOR IMPROVING CUSTOMER SATISFACTION

Improving customer satisfaction can be accomplished in two ways: 1) by changing the product (e.g., taste, odor, and level of contaminants) and/or 2) by changing consumer perception of the product. A number of recommendations for improving consumer satisfaction are put forth in the following two sections based on current practices found in the literature or in response to the "utility managers' survey."

Public Outreach

While water utilities still operate in a monopolistic market, they must now compete for consumer perception just like in an open market. Lost revenue from consumers shifting to bottled/filtered water is infinitesimal and is not a threat to utilities; less than 5% of water produced by utilities is actually consumed (AWWA 2001). However, public image and trust is at stake. In many ways, it can be argued that there is little difference between tap, bottled and filtered water in the marketplace. They are all complementary products. Advertising and "selling" public drinking water may be a new concept for water utilities, but to remain in good standing and trust, new approaches need to be taken. Some simple customer education tactics are listed below:

- Organize a public relations team and develop a response plan to deal with negative "events"
- Educate the public about source water protection and drinking water treatment.
    - Consider alternative media (television, billboards, radio).
    - Treatment plant tours/workshops offered to individuals and community groups.
- Examine changes that could be made to CCRs.
    - Consider sending them out separate from the water bill.
    - Advertise or publish them in the local paper, on television billboards and the radio to increase customer awareness
    - Include a Frequently Asked Questions (FAQ) section to address common customer complaints
    - Address local issues related to the water industry (e.g., terrorist threats, seasonal water quality problems)
- Use techniques similar to bottled and filtered water providers in the "beverage" marketplace

- Develop an advertising campaign.
- Develop a bottled water program for sale at local stores, or for distribution at charity events and related local activities.
- Develop a bottled water and/or POU/POE device service as an "added value" option for customers.
- Partner with bottled and/or filtered water companies to provide bottled water and/or filters as an "added value" option for customers.

Improving Water Quality

Some basic process changes and/or additions could be used to mitigate aesthetic problems and improve removal of microorganisms and microcontaminants. Elimination of seasonal events would likely improve customer perception of the overall quality of the tap water. However, in some cases (e.g., cooling the water during hot summer months), this would be neither feasible nor practical. Hence, the best approach in all cases would be to combine any significant process change with public education. With some changes, the water quality might be improved, but the consumer may not be aware of it (e.g., implementation of multiple disinfection barriers). In other cases, some consumers may complain even if the water quality is better simply because the water is different (i.e., not what they are "used to"). In addition to process changes and public outreach, highlighting the cost effectiveness of tap water and the benefits of a centralized treatment process would improve consumer perceptions.

FUTURE RESEARCH

Often times, research raises more questions than it answers. For this reason, an exhaustive appendix has been included with this report, which can be used beyond the analyses presented here, to further examine the area of consumer satisfaction with tap water. Specifically, an Excel™ data file has been included for those wishing to run correlations and determine relationships not presented in this report.

A few questions raised in this research which might better elucidate the tap water alternative market include the following:

1. Why is there a difference between FW and BW drinkers in terms of their motivations for selecting that type of additional treatment?
2. Is there a real health benefit (vs. a "placebo" affect) to drinking bottled or filtered water in place unfiltered tap water?
3. How much would individuals be willing to pay for bottled or filtered water service by their local utility? How would utility customers perceive such an additional service?
4. What sorts of public relations campaigns have utilities been using to successfully increase public satisfaction with drinking water? What are the associated costs?

CHAPTER 1
INTRODUCTION

BACKGROUND – TAP WATER ALTERNATIVES

Recently, much attention has been given to the consumption and use of tap water alternatives such as bottled water and POU/POE devices. Bottled water can originate from tap water, groundwater, or natural spring water and might be filtered, distilled, deionized, carbonated, or mineralized. POU/POE devices, which are used to further treat tap water, generally rest on a countertop or are installed in-line underneath a sink.

The U.S. Food and Drug Administration (FDA), which regulates the bottled water industry sets bottled water standards that are very similar to those established by the USEPA for drinking water. However, monitoring and enforcement of FDA standards, which is reserved for state agencies, is comparatively weak. According to the chief of California's FDA food safety division, "The growth of the tap water alternatives industry has...exceeded our ability to keep up..." (Oanh Ha 2001). In fact, at the national level in the FDA there exists only 1/2 a person per year devoted to bottled water regulation. The situation does not get any better for states, as only 7 of 50 states have a full time employee to monitor and enforce state bottled water regulation. (Olson 1999). A lack of resources and funding has prevented states from effectively policing bottled water. The insufficient structure and guidance in bottled water regulation has been somewhat mitigated by the presence of the International Bottled Water Association (IBWA), which represents 85% of the bottled water companies in the U.S. The IBWA has coordinated with the National Sanitation Foundation (NSF) to establish voluntary standards that are at least as stringent as those set by the FDA. An NSF seal of approval is granted to all bottled waters that meet IBWA standards.

Water produced from POU/POE devices has no specific regulatory requirements. Instead, NSF offers similar testing and certification services of POU/POE devices to those offered for bottled water companies. Currently, six National Standards Institute (ANSI)/NSF standards exist for POU/POE devices. These standards address aesthetic (e.g., taste, odor, color) and health-related (e.g., lead) issues that POU/POE devices claim to mediate. The four remaining standards are technology-specific, focusing on water softening, ultraviolet disinfection, reverse osmosis, and distillation processes. Similar to bottled water, water produced from POU/POE devices is not necessarily safer than tap water, though manufacturers may suggest that. (Hellard *et al.* 2001).

Marketing is a major component of most bottled water companies. Their willingness to spend time and money on marketing strategies is evidenced by the following advertising and expenditure plans for various companies for 2001 (Holleran 2000):

- Pepsi (Aquafina bottled water): Sponsor major sport leagues (Major League Baseball) Soccer, Professional Golf Association, National Collegiate Athletic Association); Television and printed advertisements – $10 million.
- Coca-cola (Dasani bottled water): $15 million advertising budget
- Evian bottled water: $10–15 million advertising budget

- Suntory bottled water: Acquisition of a "popular" website address (www.water.com)
- Avalon bottled water: Fantasy vacation getaway

Bottled water companies have also begun to develop accessory items that attempt to improve the convenience of tap water alternatives. For example, ChillCan has developed an aluminum pop-top can that can cool the internal liquid 30°F upon activation of the pop tab (ChillCan 2001). PopStraw (2001) likewise has advanced the bottled water market by providing a "safe and sanitary" option of drinking water through an internal straw, which emerges when the bottle is opened. Built-in handles, storage compartments, and collapsible containers are other examples of innovative accessories developed by the bottled water industry (Garrettspecialties 2001).

Marketing by the POU/POE industry focuses on consumers who desire a tap water alternative at a reduced cost. While bottled water can approach $1 per gallon, POU water from commercial sources is approximately $0.25 per gallon and even less from residential units. In addition to advertising, tap water alternative companies spend time lobbying Congress on such topics as future market regulation and the inclusion of POU devices in legislation as a government approved alternative to meet new water treatment requirements (WaterTechOnline 2001b).

CONSUMPTION TRENDS

The water market has changed dramatically in recent years. Significant growth in public/private partnerships (20–23% in the last 10 years [Dysard 1999]), intense marketing of bottled water and POU/POE devices, and the recent increase in media attention to drinking water issues has contributed to a shift in public attitudes toward tap water. A recent survey conducted by the Water Quality Association (2001), a trade organization representing POU and POE device manufacturers, indicated that 86% of Americans have concerns about their tap water. Accordingly, recent studies indicate a 10% annual increase in the use of tap water alternatives (Bottled Water Web 2000, Water Quality Association 1999). Over the past 25 years, bottled water production has experienced a 13-fold increase, with 110 new bottled water companies emerging in 2000 alone (Bottled Water Web 2000, Bachmann 2000). POU/POE devices have also gained in popularity, with nearly 2 million units being purchased annually in the U.S. and Canada (Doulton Water Filters 2001). From 1988 to 1998, annual bottled water sales tripled to about $4 billion (Suffet 2001) and the most recent figures indicate annual global bottled water revenues of $36 billion. Currently, more than 50% of Americans drink bottled water (Suffet 2001) and nearly 67% of American consumers use POU/POEs, bottled water, or both (Water Quality Association 1999).

The U.S. is not the only country experiencing increased use of tap water alternatives. Since 1995, the bottled water market in the United Kingdom has grown by 38.2% (WaterTech Online 2001a). A telephone survey showed that the percentage of people using a tap water alternative in Toronto, Canada ballooned from 4% in 1981 to approximately 40% in 1991 (Auslander and Langlois 1993). For 10 and 17% of the general population in Sydney, Australia, the primary type of water drunk at home is bottled water and filtered water, respectively (Sydney Water 2001). According to a survey conducted in 1998, 34% of consumers in the west Parisian suburb of Centre des Yvelines only drink bottled water, while 16% only drink tap water (Lyonnaise des Eaux 1998).

What is of particular interest about these trends is the fact that most people have not actually experienced tap water quality problems. In fact, only about 5% of the 55,000 water treatment systems in the U.S. ever exceed any USEPA MCL standard (EPA 2001b). Furthermore, 75% of all MCL violations are a result of monitoring/ reporting practices, and are not necessarily related to actual drinking water quality.

OBJECTIVES

Though average American tap water quality is very good, it is evident that an increasing fraction of the national population is turning to tap water alternatives as their primary drinking water source. A continuation of this trend could have serious consequences for drinking water utilities across the nation. Only by identifying and understanding the various driving forces behind this shift in water consumption trends can utilities begin to develop a strategy to compete effectively in the long-term drinking water market. Therefore, the objectives of this project were as follows:

- To measure consumer satisfaction with tap water quality.
- To investigate demographic trends in consumer satisfaction and consumption of TWAs.
- To identify the factors that cause consumers to purchase and use BW and POU/POE devices.
- To investigate whether these factors arc real or falsely perceived.
- To develop a list of recommendations for water utilities that can improve consumer satisfaction and help bridge gaps between perception and reality.

RESEARCH APPROACH

The research approach focused on identifying reasons underlying individual decisions to use alternative sources to tap water. The general approach was to identify customer attitudes and perceptions of tap, bottled, and filtered water in several geographically diverse markets. The specified approach was as follows:

1. <u>Gather physical data on each of twelve target markets across the U.S.</u>. This approach provided a physical data set to which perceptions survey data could later be associated. Utilities varied by size, type of ownership and geographic locations.

2. <u>Conduct a telephone survey to investigate public perceptions of bottled water, POU/POEs, and tap water in the target markets</u>. The survey covered pertinent issues with respect to water consumption habits and survey analyses were statistically rigorous. Approximately two hundred persons (100 TW drinkers and 100 BW or FW drinkers) were surveyed in each of the twelve target markets. Quota levels were managed in order to attain 100 respondents in each category.

3. Summarize and compare the survey results between markets and between utility management and consumers. The survey results were compared between markets (using geographical and physical data to define differences between populations) and between user groups (TW and FW/BW drinkers). These results were also compared to survey results from a utility manager's survey. In this survey, utility managers were asked to give their perception of consumer views and attitudes.

4. Formulate recommendations for water utilities. Survey results and analyses were used to develop strategies to improve public relations, CCRs, and consumer education programs.

CHAPTER 2
SURVEY METHODOLOGY

METHOD

From May 2001 to February 2002, Clearwater Research, Inc., conducted a telephone survey of household residents over 18 years of age in twelve separate water utility markets throughout the U.S. (Figure 2.1, Table 2.1). The survey was developed to collect general information about consumers' perception of, attitude toward, and use of tap, bottled, and filtered water, as well as to ask numerous questions to identify reasons underlying individual decisions to use tap water alternatives. This chapter describes the methods used to develop the survey instrument, construct the sample design, conduct the telephone survey, prepare the data for analysis, and analyze the data for significant relationships.

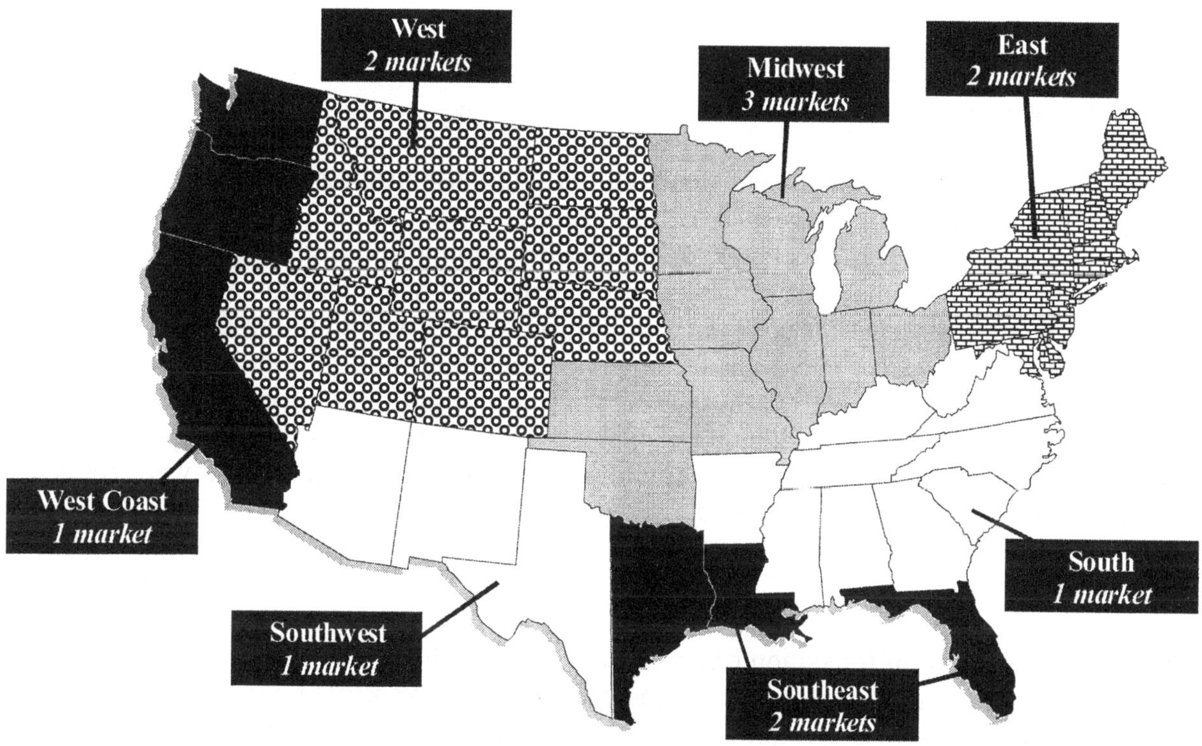

Figure 2.1 Location of the 12 water utility markets

Table 2.1
Survey dates by water utility market area

Code	Water utility market area	Start of survey	End of survey	Total days surveying
A	West 1	July 19, 2001	August 5, 2001	17
B	East 2a and 2b	July 25, 2001	January 11, 2002	170
D	Midwest 1	June 21, 2001	August 25, 2001	65
E	Southeast 1	June 21, 2001	February 1, 2002	225
F	Southeast 2	June 20, 2001	September 17, 2001	89
G	Midwest 3	May 23, 2001	July 19, 2001	57
J	Southwest	July 25, 2001	February 2, 2002	192
K	Midwest 2	May 29, 2001	July 18, 2002	50
L	West 2	May 29, 2001	July 26, 2002	58
M	East 1	July 20, 2001	January 31, 2002	195
N	South	December 6, 2001	February 1, 2002	57
O	West Coast	January 8, 2002	February 2, 2002	25

Survey Instrument

The questionnaire explored consumer attitudes and use patterns by eliciting respondent beliefs, opinions, and behaviors related to their daily consumption of TW, BW, and FW. The questionnaire instrument was developed for the non-institutionalized general adult population of the U.S. with design elements specifically incorporated for administration using CATI. The average length of the interview was 14 minutes. The survey was offered in English only.

Before executing the survey, several survey questions were evaluated by internal and external reviewers to determine whether the wording and response categories were appropriate for the survey population. In addition, a pilot test was conducted to ensure programming accuracy and effectiveness of interviewer training. Approximately 30 interviews were collected for each market during this piloting phase.

Development

Over the course of three months, the development of the survey instrument proceeded in a series of discrete stages. First, a list of candidate items was generated and loosely organized into topic-related categories for initial review. Next, intensive reviews of the item list were conducted to ensure compatibility with telephone interviewing methodology and to maintain a coherent distribution of topics. Third, a preliminary version of the instrument was constructed and evaluated using the results of the pilot testing procedures from the instrument design phase. Finally, the survey instrument was revised and administered to the target population. A copy of the finalized instrument is provided in Appendix A.

Scaling Choices

As the majority of the survey items were written specifically for this research effort, questionnaire construction was guided by the relevant professional standards of the American

Psychological Association, the National Council on Measurement in Education, and the American Educational Research Association. Response category options were specifically tailored to match the desired level of measurement and information content for each question.

Response options to all attitudinal questions were offered as a fully anchored 5-point Likert scale (i.e., "Very Satisfied," "Somewhat Satisfied," "Neither Satisfied nor Dissatisfied," "Somewhat Dissatisfied," and "Very Dissatisfied"). As many of the questions used different response categories, interviewers read all of the response options to respondents for each question. Questionnaire items that inquired about the frequency of actions, events, or behaviors were initially asked as a simple dichotomous question (i.e., "Yes" or "No") to ensure relevancy of the question to the respondent. If the response was affirmative, respondents were then asked to report the frequency of occurrence using either specific quantities (e.g., "Once," "Two to three times," or "More than three times") or general categories (e.g., "Hardly ever," "Sometimes," or "More often").

Response items for categorical questions were developed individually for each question. Items that required categorical response options fell into two general categories based on whether the response categories were read or not read to the respondent. For those questions where the response categories were read, the respondent was asked to select which category (or categories) best described his or her answer. If the order of presentation for those response lists that were read could influence a respondent's answer, then those response categories were randomly presented to the respondent.

Interviewers were trained to use the "Other" response category to capture responses that were either unanticipated or novel. These responses were collected verbatim as an open-ended response, and they were subsequently coded with the other response categories during preparation for data analysis. Interviewers were also trained to code non-responsive answers as either "I don't know / I don't understand" or "Refused" where appropriate.

"Safety" vs. "Health"

"Healthiness" and "healthfulness" both refer to the perceived health value of the water in question. Respondents were asked "Which of the following best describes how healthy the tap water you receive in your household is to drink? Would you say your water is . . .(Unhealthy / Somewhat unhealthy / Neither unhealthy nor healthy / Somewhat healthy / Healthy)" And "Considering all of these water quality features, which one is the most important to you? Would you say it was your water's. . .(Taste / Appearance / Smell / Safety /Healthfulness)."

When the survey was being developed only "safety" was originally to be queried. However it was suggested that some people may find water "safe," but still not drink it. So we came up with "safe" and "healthy" to refer to the ideas of "it's safe to drink, I won't get sick today from drinking it," and "but it's not healthy. It has contaminants in it. I think bottled or filtered water have less contaminants in them." As consumers were not provided with definitions of each, we cannot strictly say how these terms were interpreted, but the data clearly indicate consumers found them to be different questions.

Sampling

The aim of sampling is to equate unknown characteristics that may influence variation in the data and to preserve the representativeness of the sample to the target population (Henry

1990). The preferred method of ensuring the credibility and applicability of research results is to employ random sampling, where every member of a target population has a known, non-zero probability of being included in the sample. For telephone-based survey research, the typical sampling plan involves RDD of household phone numbers, where information about the distribution of residential telephone numbers is used to define the sampling frame (Cochran 1977).

Description

The sample was drawn from twelve strategically selected market areas served by a single water utility provider. These water utility providers were asked to geographically identify their service area in order to define the sample frame for each market area. Markets were identified using zip codes, service area maps, and GIS data, depending upon what the utility had available.

With the exception of the West 1 market, a random sample of telephone numbers was generated for each water utility market area. The West 1 market area could not be adequately specified geographically, so a complete phone list of residential consumers was provided by the water utility. A random sample of households was drawn from this consumer list to minimize any systematic biases introduced using this particular sampling method. For the other 11 markets, the sample was generated using a random digit dialing routine. Before execution of the survey nonresidential phone numbers were purged by using a combination of line signal checks (for fax numbers) and directory searches (for business / commercial listings).

Quota Cells

The sampling plan required collecting approximately 200 interviews per market for a total sample size of 2,400. To ensure that respondents in each market contained respondents who were either (1) primarily TW drinkers or (2) primarily BW or FW drinkers, two quota cells were established for each group. The survey was managed such that in each market approximately 100 TW drinkers and 100 TWA drinkers (BW or FW) were interviewed (Table 2.2). A screening question was used to identify respondent's eligibility based on drinking water preference while at home.

Table 2.2
Water utility market area population information

Water utility market area	Total households	Total completed interviews
West 1	77,850	200
East 2a and 2b	379,100	211
Midwest 1	181,612	200
Southeast 1	285,200	195
Southeast 2	212,400	200
Midwest 3	17,773	197
Southwest	188,500	159
Midwest 2	10,319	197
West 2	148,690	200
East 1	1,388,700	200
South	51,100	185
West Coast	79,500	123

In cases where the sample size for one group was met (i.e., 100 respondents), the screening information was still collected from every respondent. Thereafter this information may be used to estimate the overall incidence of TW and TWA drinkers in the market. Table 2.3 presents the estimated incidence of these groups in the population of interest.

Table 2.3
Incidence of tap water and tap water alternatives drinkers

Water utility market area	Total TW drinkers	Total TWA drinkers	Projected incidence of TW drinkers	Projected incidence of TWA drinkers
West 1	136	100	58%	42%
East 2a and 2b	223	111	67%	33%
Midwest 1	137	100	58%	42%
Southeast 1	95	371	20%	80%
Southeast 2	210	100	68%	32%
Midwest 3	381	96	80%	20%
Southwest	59	282	17%	83%
Midwest 2	119	97	55%	45%
West 2	185	99	65%	35%
East 1	197	100	66%	34%
South	127	85	60%	40%
West Coast	22	142	13%	87%

Error Estimates

A sample size of 200 for each water utility market area was determined to be sufficient for the accuracy of the survey results. The overall planned sample size was 2400 completed interviews, but the actual total sample size achieved was 2268. The obtained sample size was lower than originally intended given the constraints of the fielding period and the incidence rate of TW and TWA drinkers for each market.

The overall sample size was projected to achieve a 95% confidence interval of ±2% about binomial point estimates for all twelve markets combined (see Table 2.4). The sample size for each market utility area was projected to achieve a 95% confidence interval of ±7–8%. Estimates projecting to all TW drinkers or TWA drinkers across all twelve market areas combined were projected to achieve a 95% confidence interval of ±3%.

Table 2.4
Error estimates for individual and combined markets

Water utility market area	Completed interviews	Estimated confidence intervals	TW drinkers	TWA drinkers
West 1	200	(±7%)	100	100
East 2a and 2b	211	(±7%)	100	111
Midwest 1	200	(±7%)	100	100
Southeast 1	195	(±7%)	95	100
Southeast 2	200	(±7%)	100	100
Midwest 3	197	(±7%)	101	96
Southwest	159	(±8%)	59	100
Midwest 2	197	(±7%)	100	97
West 2	200	(±7%)	101	99
East 1	200	(±7%)	100	100
South	186	(±7%)	101	85
West Coast	123	(±8%)	22	101
	2268		1079	1189
		(±2%)	(±3%)	(±3%)

Interviewing Respondents

The CATI system managed and monitored calling, controlled distribution of telephone numbers to interviewers, and consolidated the collected survey data. When necessary, interviewers left voice mail messages briefly describing the study's purpose and asking respondents to call back at their convenience. The complete script of the voice message is provided in the text of the questionnaire instrument in Appendix A. Surveys were administered primarily during evening and weekend hours, although some calls were made during the weekday if no contact was made at other times. Each phone number was called until an interview had been completed, the respondent refused to take the survey, or the phone number had been called 8 separate times without an answer.

A set of internal codes adopted from the CDC BRFSS protocol was used to track response rate, cooperation rate, and refusal rate. The input variables for these rates and the corresponding formulas used to calculate the rates are given in Appendix B (variables) and C (formulas). Calculated rates for each of the 12 markets can be found in Appendix D.

The MRA maintains a database available via the Internet to track survey performance for the marketing research community. In addition to the survey performance information for the present study, Table 2.5 presents industry-wide survey performance statistics from the MRA database for all RDD phone surveys and for phone surveys that average 15 minutes. Data collection for this study performed remarkably similar to other studies in terms of response rates and cooperation rates, but surpassed the performance of most other MRA studies in terms of overall refusal rate.

Table 2.5
Comparison of survey performance rates

Market	Response rate (%)	Cooperation rate (%)	Refusal rate (%)
Southeast 1	5	9	28
Southeast 2	8	12	16
East 1	9	13	31
Midwest 3	9	13	29
East 2	9	13	28
Midwest 2	10	14	26
South	10	14	44
West Coast	10	16	30
West 2	11	17	27
Southeast 2	11	19	22
Midwest 1	12	17	27
West 1	15	22	47
National average	10	15	30
Overall RDD surveys	13	16	38
15 RDD surveys	10	11	79

Data Preparation & Analysis

Once an interview was complete, data were exported from the CATI system into a statistical program (SPSS) used for data analysis. Open-ended responses where respondents could mention an option not listed in the original survey categories were examined and relocated where appropriate. For example, if responses such as "bad", "terrible", "awful", or "a dead animal" were given for a question about the smell of tap water, one categorical descriptor (e.g., "bad") would be used to include all these related, negative descriptions. Once data had been cleared and exported to SPSS, general statistics (e.g., frequencies, mean, standard deviations) were run.

Using findings from the initial output of results and an intuitive feel for likely relationships, correlations and cross-tabulations were expected for a number of survey questions.

For any 2 questions where responses could categorize using an ordinal scale (e.g., 1 = very dissatisfied, 3 = neutral, 5 = very satisfied), t-tests, ANOVA statistics, and multiple regression analyses were used. In cases where categorical responses were not scalable (e.g., responses for water smell-musty/earthy, chlorinous, fresh, stale), chi-square statistics were used to make general comparison among different sub-populations. All tests for difference/significance were conducted with a α-value of 5% (i.e., 95% level of confidence).

Quality Control

Collection and analysis of data was cross-checked throughout the survey. Preliminary testing allowed the survey instrument to be refined in terms of wording and logic flow. During data collection, interviewers received constructive feed-back from trainers. After data collection, statistical tests (t-test, ANOVA) and input data (values and calculated frequencies) were analyzed and reanalyzed by at least 2 analysts.

Interviewers were also supplied with the local water utility contact information for respondents who wanted to verify the legitimacy of the survey. Interviewers were regularly monitored and reviewed by trainers who have at least 12 months interviewing experience. Using NetOp software provided trainers the ability to monitor interviewers in session, make changes to answers, and send comments/suggestions to the interviewer. Finally, quality check notes are given to each interviewer at the end of a session of interviews. These notes included statistics regarding the ratio of refused to completed interviews, number of completed interviews per hour, and number of dialings per hour. Thus, interviewers receive training before, during, and after each session.

Supervisors overseeing interviewers all had at least 12 months interviewing experience. Special training related to the project was also given to supervisors so that they could answer any project specific questions that arose during the interviews. Multiple supervisors were used to monitor the 90 different CATI stations used during the survey.

Manager's Survey

In addition to the consumer survey, a manager's survey was sent out to approximately one-hundred water utilities (from approximately one-hundred thirty initial telephone and email contacts) including all participating utilities and any other utilities expressing interest. One hundred responses were collected. Survey questions roughly followed the same format as for the consumer survey; the first set of questions focused on tap water quality, the second set focused on utility service. Utility managers were asked to respond to questions from the viewpoint of their customers. For example, in the consumer survey, the question would be asked, " What is your overall satisfaction with the quality of tap water in your home?" However, a utility manager would receive the question, "What is your average customer's level of satisfaction with the overall quality of tap water in their household?" Surveys were e-mailed or faxed and results hand-entered into Excel™ for further statistical analysis.

CHAPTER 3
NATIONAL MARKET RESULTS

Twelve markets across the United States were surveyed via telephone. Approximately 200 persons were interviewed in each region: 100 TW and 100 TWA drinkers. Markets varied in terms of geographic location, size, and type of ownership. The aggregate results of this survey from all 12 markets represents a snap shot of the national water consumer market. These data will be discussed in three sections examining consumer perception of water quality, utility service, and motivators for switching to tap water alternatives. Key survey questions were selected to highlight TW and TWA drinkers' opinions. A complete list of the results, overall and by market, is provided on the accompanying CD-ROM.

CONSUMER PERCEPTION OF WATER QUALITY

Geographic location and drinking water preference (TW vs. TWA drinkers) impacted consumer satisfaction levels. Comparisons between different markets and consumer groups were facilitated by the use of group averages. Questions regarding satisfaction levels used a 5 point grading system; "1" was equivalent to "very dissatisfied" while "5" was equivalent to "very satisfied".

Satisfaction with Tap Water Quality

People were asked several questions focusing on overall water quality, safety, and healthiness and confidence that their tap water meets or exceeds state and federal requirements. Questions were phrased as follows with specific water quality feature inserted for "WQ feature."

- Which of the following best describes your satisfaction with the "WQ feature" of the tap water you receive in your household. (1. Very dissatisfied, 2. Dissatisfied, E. Neither dissatisfied nor satisfied, 4. Satisfied, 5. Very Satisfied) (Q 005, Q025, Q026, Q110)?

For each market an average response was determined for the two consumer groups, TW and TWA drinkers (Table 3.1). On average, TW drinkers were more satisfied than TWA drinkers in areas of tap water quality, safety, and healthiness in each market. Differences were statistically significant in all but four cases.

Table 3.1
Regional analysis of the effect of water consumer group on perceptions of tap water

Geographic region	Market	Overall satisfaction with TW quality		Description of TW safety		Description of TW healthiness		Confidence in TW safety	
		TW	TWA	TW	TWA	TW	TWA	TW	TWA
West Coast	West Coast	3.5*	3.2*	4.3	3.4	4.4	3.3	3.6*	3.1*
Southwest	Southwest	3.6	2.8	4.4	3.7	4.2	3.1	4.0	3.1
West	West 1	3.9	3.6	4.6	4.3	4.4*	3.8*	3.9	3.7
	West 2	4.2	3.7	4.7	4.2	4.5	4.0	4.1	3.5
Midwest	Midwest 1	4.0	3.4	4.6	4.0	4.4	3.8	4.0	3.3
	Midwest 2	4.3	3.6	4.8	4.3	4.7	4.0	4.4	3.7
	Midwest 3	4.3	3.8	4.8	4.4	4.6	4.1	4.2	3.6
East Coast	East 1	3.8	3.2	4.5	4.0	4.1	3.6	3.8	3.1
	East 2a and 2b	4.1	3.5	4.7	4.1	4.6	3.7	4.1	3.3
Southeast	Southeast 1	3.7	3.3	4.5	4.0	4.3	3.5	3.7*	3.5*
	Southeast 2	3.9	3.2	4.5	4.0	4.3	3.6	3.7	3.4
South	South	4.3	3.6	4.8	4.4	4.6	3.8	4.4	3.7

TW Tap water (drinkers)
TWA Tap water alternative (drinkers...bottled or filtered water)
* Indicates that no statistically significant difference was measured between average responses for TW and TWA drinkers in that market

Though TWA drinkers were less satisfied than TW drinkers, their satisfaction averages were above 3.0 in every case except one (Southwest market). Therefore, while some geographic regions have experienced considerable shifts (i.e., >50%) in the number of consumers using tap water alternatives, average tap water satisfaction among the TWA drinkers was still above "neutral".

A combined market analysis of the same water quality issues is presented in Table 3.2. For this comparison, TWA drinkers are separated into those who primarily drink filtered water at home and those who primarily drink bottled water. TW, FW, and BW drinkers' satisfaction levels decreased in corresponding order.

Table 3.2
Effect of specific water consumer group on perceptions of tap water

Water quality issue	Average response		
	TW	FW	BW
Overall satisfaction with TW quality	4.0	3.6	3.1
Description of TW safety	4.6	4.3	3.7
Description of TW healthiness	4.4	4.0	3.4
Confidence in TW safety	4.0	3.6	3.1

TW tap water (drinker)
FW filtered water (drinker)
BW bottled water (drinker)

TW and TWA drinkers' perceptions of various tap water quality features are illustrated in Figures 3.1 and 3.2. The highest level of consumer satisfaction was associated with the safety of tap water (92% and 73% for TW and TWA drinkers, respectively) while the lowest level of consumer satisfaction was associated with taste (81% and 52% for TW and TWA drinkers, respectively). While TWA drinkers were equally satisfied with tap water safety and appearance, there was a higher level of satisfaction regarding safety. Fifty-four percent of TWA drinkers were "very satisfied" with safety as compared to 18% "very satisfied" consumers regarding tap water appearance. An interesting feature was the percentage of TWA drinkers not having a distinct satisfaction level. In all water quality satisfaction questions (Fig. 3.1, 3.2, 3.5, 3.6), significantly more TWA drinkers responded neutrally (i.e., 3.0) as compared to TW drinkers. This finding suggests confusion or indifference to water quality issues but would require additional study to verify the extent of this neutrality.

Among TW drinkers, less than 10% were dissatisfied or very dissatisfied with any of the water quality features cited (overall quality, taste, appearance, smell, safety, and healthiness). Between 10 and 35% of TWA drinkers were dissatisfied with their tap water's appearance, smell, or safety. Given that the odor threshold concentration for many compounds such as free chlorine is much lower than its taste threshold concentration, it is somewhat surprising that consumers were less satisfied with taste than with odor. It is possible that consumers are confusing what they smell with what they taste, as the flavor of any substance includes a significant smell component.

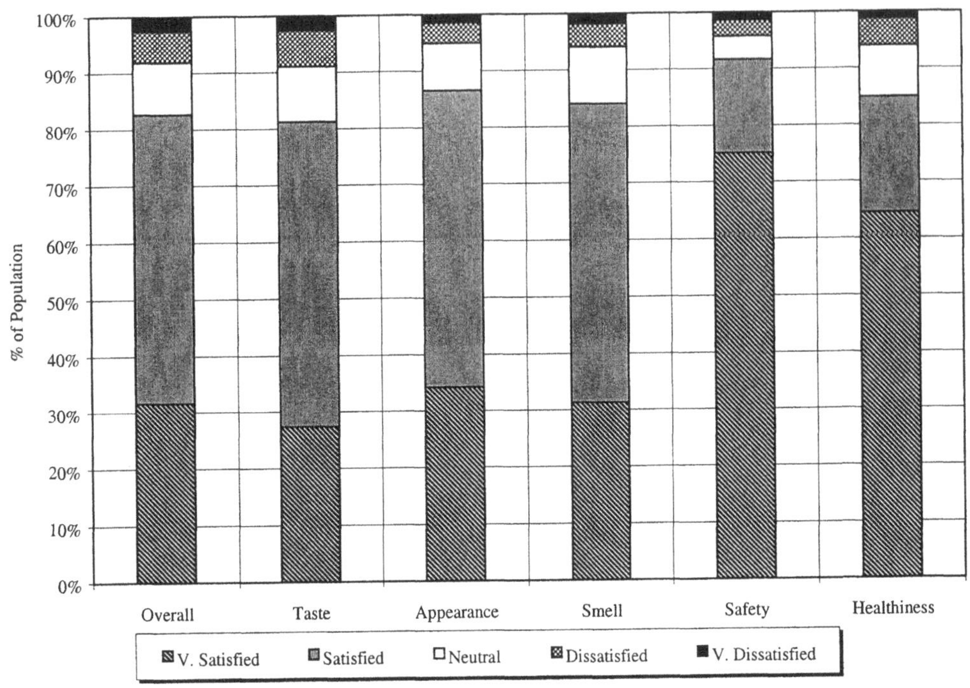

Figure 3.1 TW drinkers perceptions of tap water quality features on the national level (Q005, Q010, Q015, Q020, Q025, Q026)

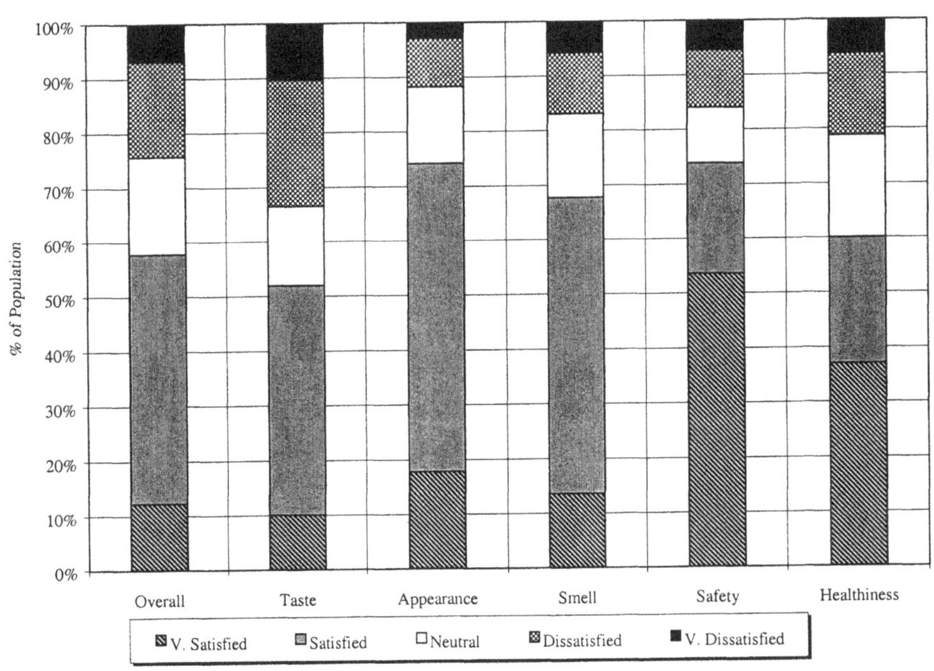

Figure 3.2 TWA drinkers perception of tap water quality features on the national level (Q005, Q010, Q015, Q020, Q025, Q026)

Important Water Quality Features

Consumers were asked to choose the most important feature of water quality from the following list of attributes: taste, appearance, smell, safety, or healthfulness. A comparison of responses among TW, FW, and BW drinkers (combined markets) is presented in Figure 3.3. There was very little difference among consumer groups as to their choice of most important water quality feature. Nearly half of the TW, FW, and BW drinkers surveyed indicated that safety was most important. Taste and healthfulness were the only other water features chosen with any frequency among those surveyed.

Consumers were asked to describe the taste of their local tap water. Responses are illustrated in Figure 3.4. Sixty-eight percent of TW drinkers and 46% of TWA drinkers described the taste of the tap water as good (i.e., fresh, okay, or no perceived taste). Eighteen percent of TWA drinkers and 5% of TW drinkers indicated that their tap water has a chlorinous flavor. A low percentage of both groups described their tap water taste as metallic or musty/earthy (≤ 5%), while a portion of both TW and TWA drinkers listed their water taste as "Other" (22 and 24%, respectively). This "Other" category included responses that ranged from "nasty" to "heavenly".

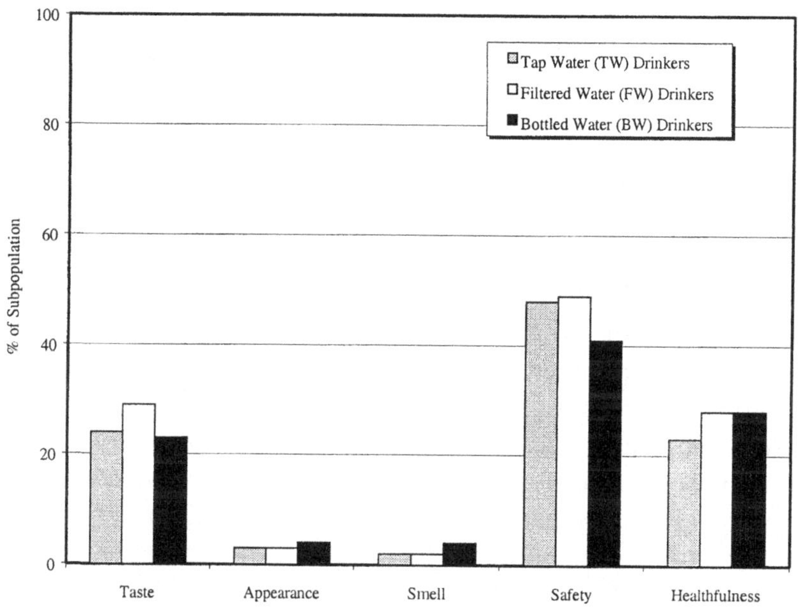

Figure 3.3 Most important water quality feature for each consumer group (Q 030)

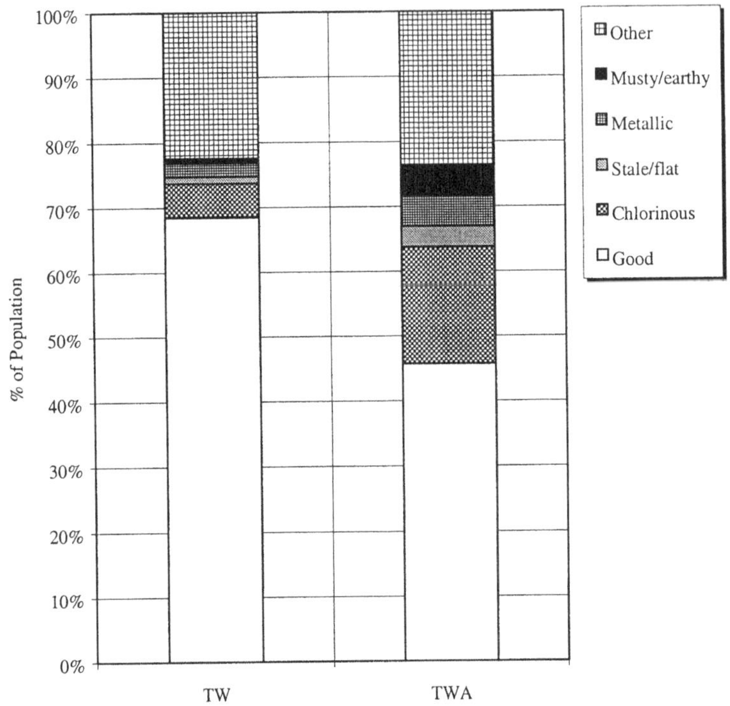

Figure 3.4 Description of tap water taste as perceived by TW and TWA drinkers at the national market level (Q 045)

CONSUMER PERCEPTION OF UTILITY QUALITY

TW and TWA drinkers were asked to describe their overall level of satisfaction with their local water utility. Overall, TW and TWA drinkers were satisfied with their local water utility; only 3% of TW drinkers and 14% of TWA drinkers were dissatisfied. These data suggest that most TWA drinkers view filtered or bottled water as a luxury item, not something that must be purchased due to shortcomings of their water utility. With respect to complaint resolution, TW drinkers and TWA drinkers were equally dissatisfied (20 and 23%, respectively).

Large and small utilities have been grappling with the ever-growing need to supply their customers with water of higher and higher purity. All of the participating utilities met or exceeded state and federal requirements, but when their customers were polled, an average of 6 and 16% of TW and TWA drinkers, respectively, were not confident this statement was true (Figure 3.5 and 3.6). These percentages were as high as 22% and 29% for TW and TWA drinkers, respectively, in some markets. As discussed previously, a portion (10%) of TWA drinkers gave a neutral response for this question. This finding, along with an overall lower level of confidence as compared to TW drinkers suggested that while TWA drinkers were somewhat satisfied with tap water quality, there was an underlying concern about basic safety and confidence concerning their tap water and local utility. This concern for safety also appeared again when TWA drinkers were asked why they use tap water alternatives (see "MOTIVATION FOR DRINKING TWAs" later in this chapter).

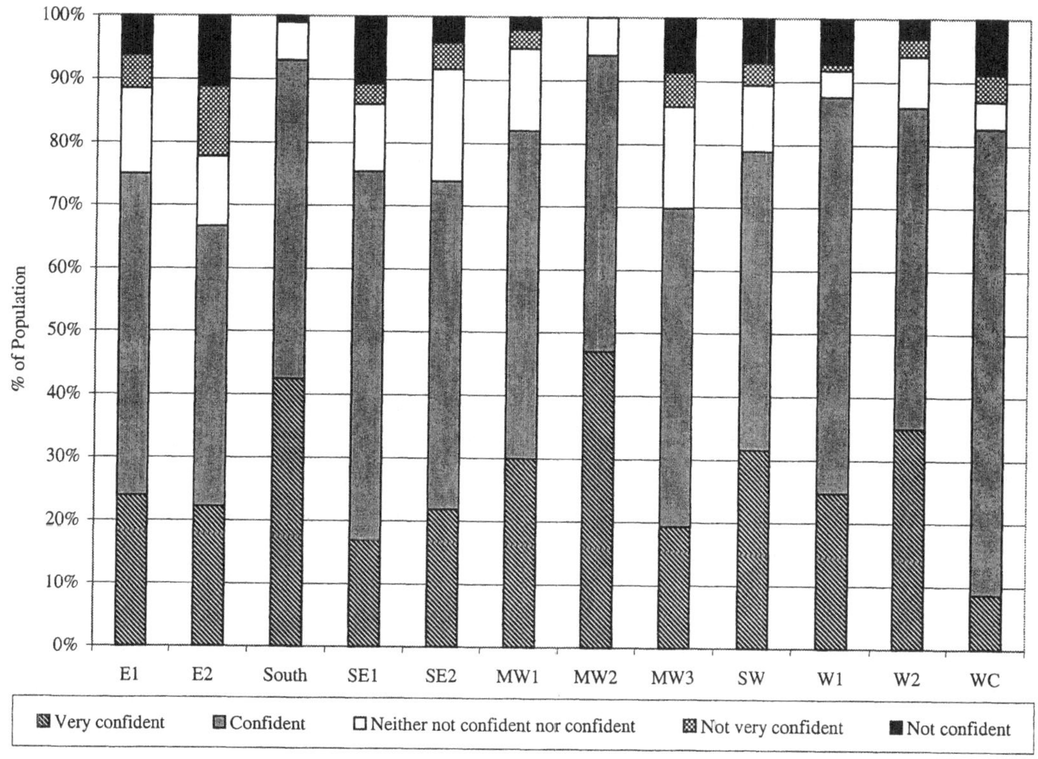

Figure 3.5 TW drinkers' confidence that their household tap water is safe (Q110)

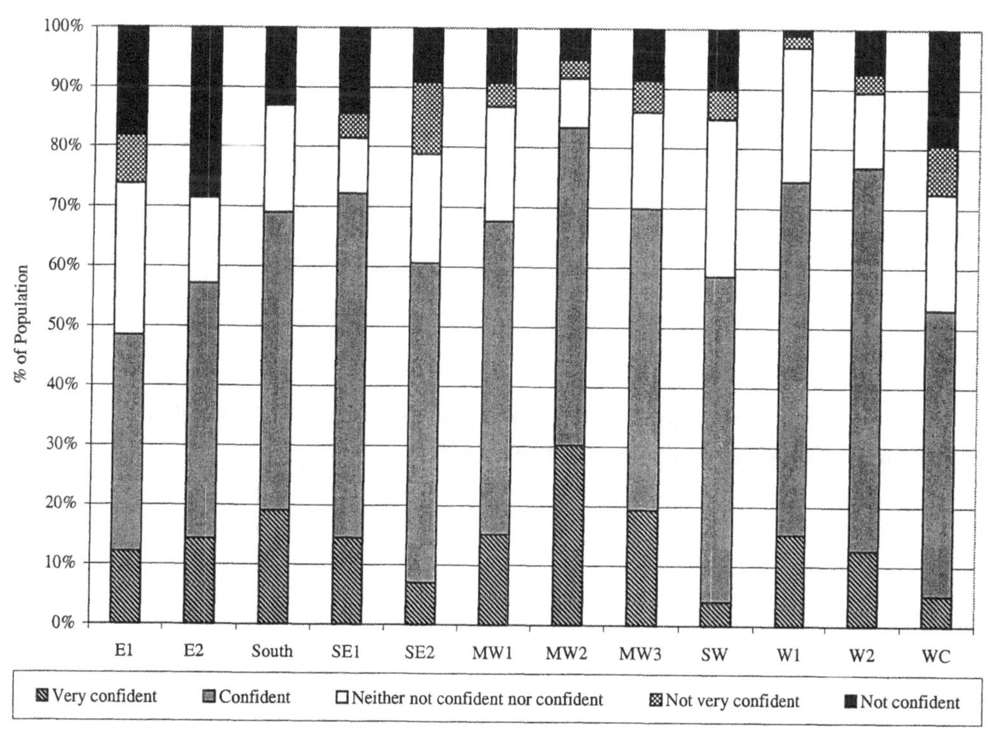

Figure 3.6 TWA drinkers' confidence that their household tap water is safe (Q110)

MOTIVATION FOR DRINKING TAP WATER ALTERNATIVES

TWA drinkers were asked to identify the most important reason they used tap water alternatives. The results are illustrated in Figure 3.7. Sixty-eight percent of FW drinkers indicated that tap water safety was their primary reason for drinking filtered water (they were not offered healthiness as an option). BW drinkers cited taste, safety, and healthiness in equal proportions as the primary reasons they drink bottled water (30, 28, and 28%, respectively). The selection of TWAs seems to be primarily driven by perceptions of increased safety, followed distantly by aesthetics (taste).

A regional analysis showed the same trend as observed for the national average. FW drinkers cited "safety" while BW drinkers cited "taste," "safety," and "healthiness" as their primary reasons for using tap water alternatives (Table 3.3, 3.4). Sample sizes for each market varied between 12 (Midwest 3) and 45 (West Coast), which stresses the fact that quota cells are necessary in order to provide a large enough population from which one can make conclusions with a reasonable level of confidence.

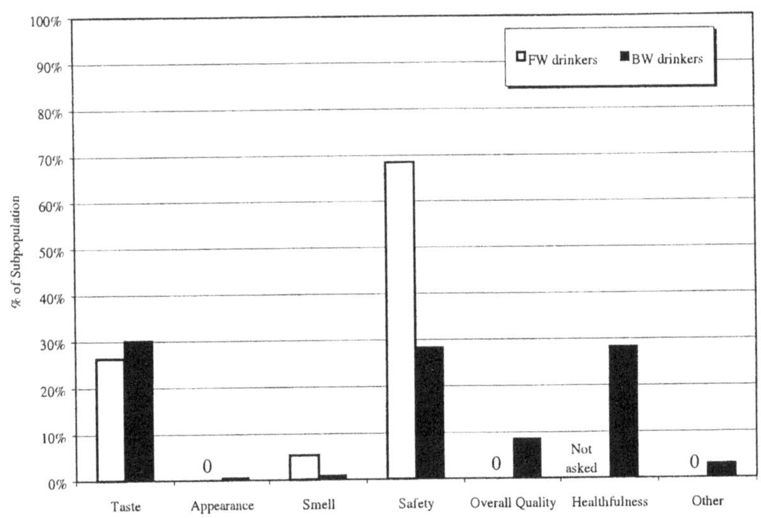

Figure 3.7 Reasons for drinking bottled and filtered water as ranked by FW and BW drinkers on the national market level (Q091, Q102)

Table 3.3
Analysis of primary reason for drinking bottled water (Q 102)

Geographic region	Market	Responses (% of total)							Number of responses
		Taste	Appearance	Smell	Safety	Overall quality	Healthiness	Other	
West Coast	West Coast	38	0	2	24	7	27	2	45
Southwest	Southwest	40	0	0	25	5	25	5	33
West	West 1	29	0	0	38	8	13	12	24
	West 2	42	0	0	23	8	19	8	20
Midwest	Midwest 1	21	3	0	38	13	25	0	24
	Midwest 2	31	3	0	22	14	28	2	30
	Midwest 3	42	0	8	25	0	25	0	12
East Coast	East 1	19	0	0	31	11	36	3	36
	East 2a and 2b	33	0	0	33	5	29	0	24
Southeast	Southeast 1	31	0	3	23	8	35	0	26
	Southeast 2	23	0	0	27	5	40	5	22
South	South	19	0	0	33	16	19	13	31
Combined markets		30	<1	<1	28	9	28	4	327

Table 3.4
Analysis of primary reason for drinking filtered water (Q 091)

Geographic region	Market	Responses (% of total)							Number of reponses
		Taste	Appearance	Smell	Safety	Overall quality	Healthiness	Other	
West Coast	West Coast	31	4	4	57	4	0	0	26
Southwest	Southwest	16	0	0	68	12	0	4	25
West	West 1	57	0	0	32	11	0	0	28
	West 2	20	0	0	70	10	0	0	10
Midwest	Midwest 1	40	0	0	55	5	0	0	20
	Midwest 2	33	0	6	50	11	0	0	18
	Midwest 3	18	0	0	53	29	0	0	17
East Coast	East 1	11	0	0	74	15	0	0	19
	East 2a and 2b	32	0	0	55	9	0	4	22
Southeast	Southeast 1	27	0	0	55	18	0	0	22
	Southeast 2	25	0	0	63	12	0	0	8
South	South	27	0	0	57	13	0	3	30
Combined markets		29	<1	0	57	11	0	1	245

CHAPTER 4
TARGET MARKET CHARACTERIZATION

UTILITY OVERVIEW

Water quality data from 12 market areas across the U.S. were collected and examined before execution of the survey. A summary of utility characteristics and water qualities is presented in Tables 4.1 and 4.2, respectively. Utilities varied in geographic location, size, and type of ownership (private versus public).

A brief description of each utility is summarized under the paragraph entitled "Physical Data." General results of the consumer survey by target market are then presented under two sections: 1) Consumer Perception of Water Quality, which focuses on consumer perception of tap water quality and reasons for drinking other types of water, and 2) Consumer Perception of Overall Utility Quality, which focuses on consumer satisfaction with their local water utility provider. In most cases, the number of consumers registering a complaint was very small, and so the data is not statistically significant. As with Chapter 3, only a subset of the survey questions was selected for this discussion. The full list of market and national data is detailed in Appendix F.

West Coast Market

Physical Data

This West Coast market is part of a large private company that supplies 165 million gallons of water per day to consumers throughout the state of California. This specific utility has a supply capacity of 8 mgd; less than 10% of the supply comes from local wells. The remaining water supply is purchased and comes from various surface waters located throughout the state.

Blending local well water with purchased surface water results in a blended water of moderate hardness (161 mg/L as $CaCO_3$) and TDS (388 mg/L). Blending helps reduce what otherwise would be a very hard (484 mg/L) and high TDS (1,383 mg/L) groundwater.

Table 4.1
General summary of water utilities in each of the 12 target markets

Geographical Region	Market	Utility capacity* (mgd)	Ownership status	Potential biasing media reports †	Aesthetic issues‡.
West Coast	West Coast	8 (GW)	Private	*Erin Brokovich* (Universal Studios 2000) Headlines on MTBE contamination of California wells	
Southwest	Southwest	105 (SW)	Public	"East Valley's Supply is Safe, It just Tastes Terrible" (Yantis 2001)	T&O
West	West 1	80 (GW) 16 (SW)	Private		sediments
West	West 2	50 (SW)	Public		T&O, sediments, color
Midwest	Midwest 1	165 (SW)	Public	"Kansas City's Bottled Water Experiences" (Hall and Miller 2001)	
Midwest	Midwest 2	13 (SW)	Public	*Cryptosporidium* outbreak in Milwaukee, WI (CDC 2002)	T&O§
Midwest	Midwest 3	48 (SW)	Public		T&O, color
East Coast	East 1	300 (SW)	Public	"Philadelphia water bad for elderly's health" (Gevirtz 1999) "A different danger–lead in the water–stalks Phila. Schools" (Ferrick 2000)	T&O
East Coast	East 2a	48 (GW, SW)	Private	"Contaminated water found at BHC Company" (McKinney, 2000)	T&O – earthy/musty, sediments
	East 2b	102 (SW) 16 (GW)	Public	"Chemical spill could threaten water supply" (Viviano 1998)	
Southeast	Southeast 1	40 (GW)	Public	"Worst Drought in History Leaves Florida High and Dry" (Newsmax.com 2001)	T&O
Southeast	Southeast 2	80 (SW)	Public	"City's murky tap water too much for most on council" (Suydam 1999)	Color, appearance
South	South	6 (GW)	Public		Temperature Addition of RO greatly improved water quality in 1991

* This represents total system capacity, many include supplies from multiple water treatment plants and/or wells.

† These events detail media reports and related publicity concerning water quality and safety that may have influenced consumers. **These reports do not reflect any failure of the utility concerned to meet Federal drinking water standards.**

‡ Many of the reported aesthetic issues are seasonal and/or restricted to certain areas of the distribution system.

§ Taste and odor was not an issue at the time of the survey.

Table 4.2
Average treated water qualities for utilities participating in this survey[§]

Market	Temp. range (°C)	PH	Alkalinity (mg/L $CaCO_3$)	TDS (mg/L)	Hardness (mg/L $CaCO_3$)	Residual disinfectant
West Coast	5-28	7.5	98	388	161	Chlorine
Southwest	14 - 27	7.6	141	579	221	Chlorine
West 1	14 - 29	7.4* 6.1-8.7†	46 78	53 36-551	35 15-241	Chlorine
West 2	3 - 15	8.2	28	53	41	Chlorine
Midwest 1	5 - 29	9.5	51	373	136	Chloramines
Midwest 2	1 - 24	8.2	110	158	136	Chlorine
Midwest 3	8 - 18	8.4	65	111	101	Chlorine
East 1	1 - 34	7.5	40 – 70	200 – 300	90 - 150	Chloramines
East 2a East 2b	2 - 27 1 - 28	7.4 7.3	54 12	116 138	48 70	Chlorine
Southeast 1	25 – 28	8.0	127	183	140	Chlorine
Southeast 2	11 - 25	7.8	141	480	182	Chloramines
South	18-32	8.3	28; 138‡	110; 268	55; 33	Chloramines

* Surface water

† Ground water (a range of values is given in most cases)

‡ First value indicates purchased water quality; second value indicates local treated water quality Up to 25% of total supply capacity may be purchased.

§ Ranges are given in cases of more than one treated water source (e.g., purchased water, water from more than one WTP/wellhead)

Consumer Perception of Water Quality

Perceptions of the West Coast market TW and TWA drinkers of their tap water quality are illustrated in Figures 4.1 and 4.2. Regarding satisfaction with the overall quality of tap water, 73% of TW drinkers were satisfied to very satisfied, as compared to 47% of TWA drinkers. In both groups of consumers, tap water appearance was perceived to be satisfactory to very satisfactory (82% for TW drinkers and 67% for TWA drinkers). The smallest proportion of satisfied to very satisfied consumers was observed for tap water taste (67% TW drinkers and 35% TWA drinkers). A small number of consumers were very dissatisfied with overall tap water quality features (5–14% for TW drinkers as compared to 3–12% for TWA drinkers).

A comparison between all three user groups (TW, BW, and FW drinkers) showed a consistent trend regarding satisfaction levels. TW drinkers were more satisfied for each tap water quality feature followed by FW drinkers, and finally BW drinkers. "Average" satisfaction levels were calculated for each group in order to make this comparison.

BW and FW drinkers were asked to cite the most important reason they used their particular tap water alternative. The results are illustrated in Figure 4.3. Out of the overall tap water quality features of taste, appearance, smell, safety, and overall quality, tap water safety ranked as the number one reason for FW drinkers (58%) followed by taste (31%), overall quality (4%), smell (4%), and appearance (4%). Out of the overall tap water quality features of taste, appearance, smell, healthiness, safety, and overall quality, tap water taste (38%) ranked as the number one reason for BW drinkers followed by safety (24%), overall quality (7%), smell (2%) and healthiness (2%). These observations match the TWA group's perceptions of tap water features.

The taste of tap water, as perceived by TW and TWA drinkers, is shown in Figure 4.4. Thirty-five percent of TW drinkers described the taste of tap water as good versus 28% of TWA drinkers. Only 10% of TWA drinkers perceived tap water as having chlorinous flavors versus 5% of TW drinkers.

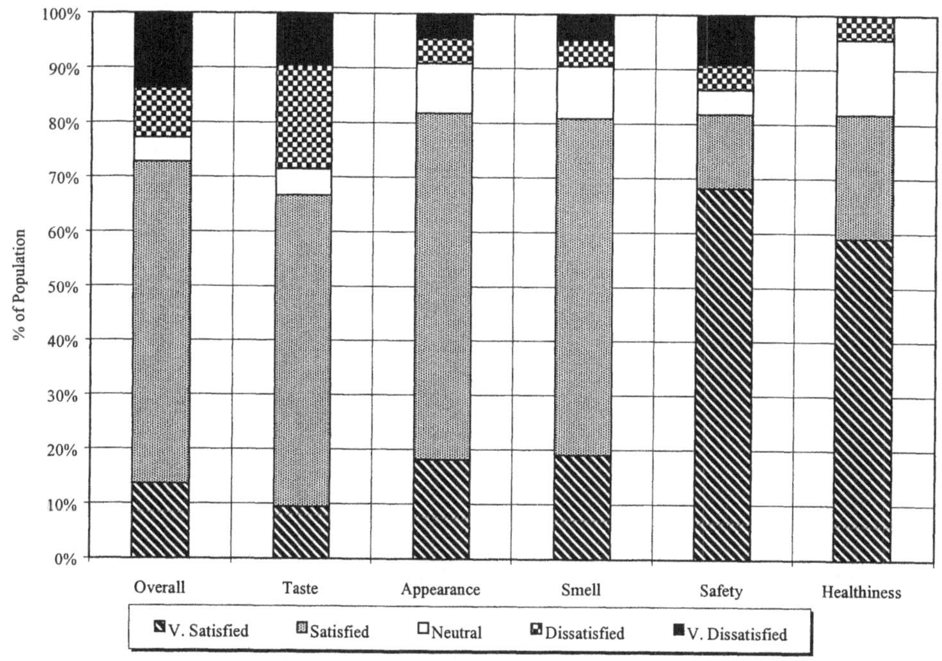

Figure 4.1 TW drinkers' satisfaction levels for various tap water quality issues - West Coast market (Q005, Q010, Q015, Q020, Q025, Q026)

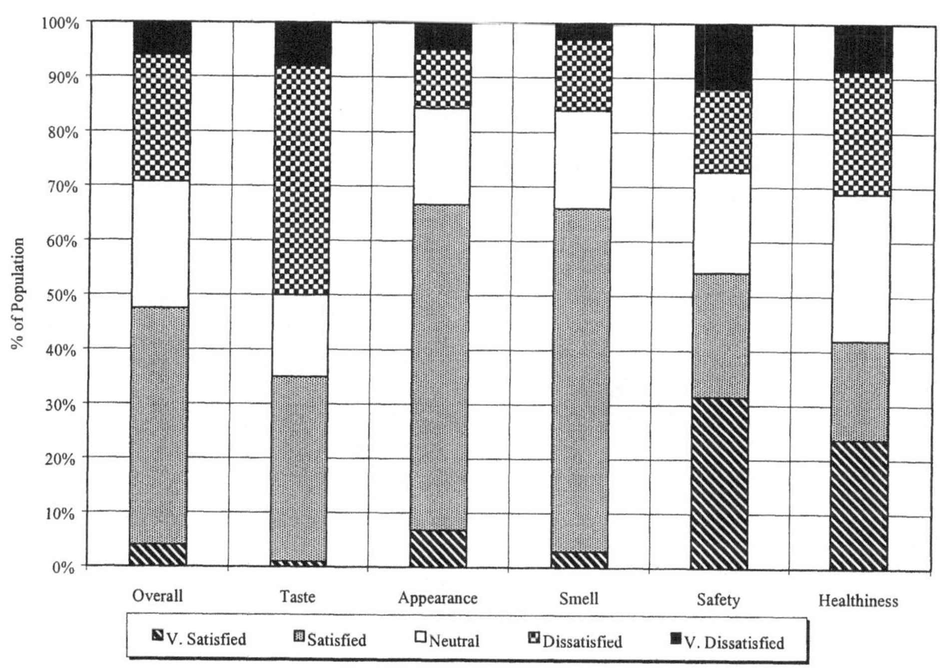

Figure 4.2 TWA drinkers' satisfaction levels for various tap water quality issues - West Coast market (Q005, Q010, Q015, Q020, Q025, Q026)

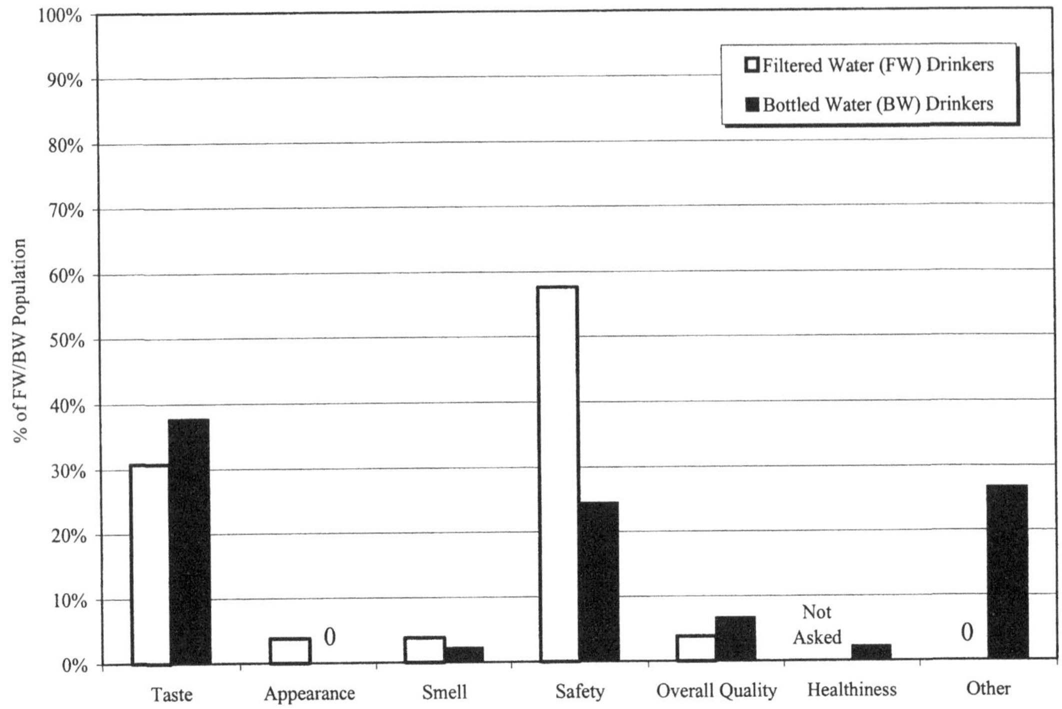

Figure 4.3 Most important reasons for drinking tap water alternatives as cited by BW and FW drinkers - West Coast market (Q091, Q102)

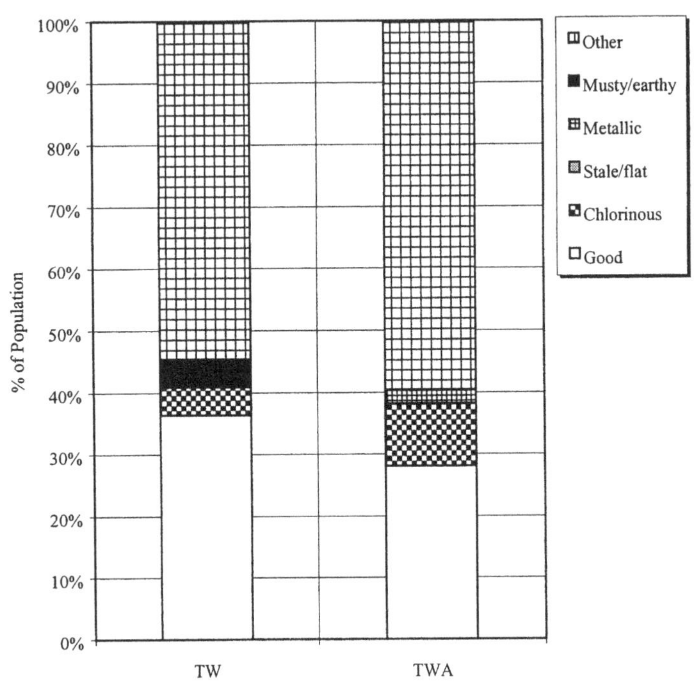

Figure 4.4 Description of tap water taste as perceived by TW and TWA drinkers - West Coast market (Q045)

Consumer Perception of Overall Utility Quality

West Coast market consumers from both consumer groups described their overall satisfaction with their local water utility. Eighty-seven percent of TW drinkers were satisfied with their local water utility as compared to 55% of TWA drinkers. No responses were collected from TW drinkers regarding their satisfaction with their provider's record for resolution of complaints. Out of the four responses received from TWA group, half were satisfied with the complaint resolution of their local provider and the other half had no opinion on the matter.

Southwest Market

Physical Data

The Southwest market surveyed during this project is a desert community that relies on transported canal water. Two water treatment plants supply the 50 million gallons needed daily by the Southwest market. The majority of water is taken from the Salt River and Central Arizona Project water, with local wells reserved for use during times of shortage. The long transport time of canal water, with little protection from the sun, can result in microbial growth that leads to taste and odor problems. When this occurs, the Southwest market supplements the treatment process with powder activated carbon. Mitigating these seasonal episodes of geosmin and MIB cost the utility approximately $700,000 a year.

Water quality varies slightly throughout the year. Typical for waters running through arid regions, source water is high in dissolved salts, specifically alkalinity and hardness. Seasonal variation in organic carbon results in a peak of about to 3 mg/L TOC during the summer.

Consumer Perception of Water Quality

Perceptions of the Southwest market TW drinkers and TWA drinkers of their TW quality features are illustrated in Figures 4.5 and 4.6. Regarding satisfaction with the overall quality of tap water, 66% of TW drinkers were satisfied to very satisfied, as compared to only 36% of TWA drinkers. Safety was rated the highest by or both consumer groups (85% for TW drinkers and 64% for TWA drinkers). Fewer satisfied were observed in the taste feature of tap water for the TWA group (34% as compared to 61% for TW group). A high number of TWA drinkers were very dissatisfied with taste (23%) and overall quality (15%) of tap water.

BW and FW drinkers were asked to cite the most important reason they used tap water alternatives. The results are illustrated in Figure 4.7. Out of the overall tap water quality features of taste, appearance, smell, safety, and overall quality, tap water safety ranked as the number one reason for FW drinkers (68%) followed by taste (16%), overall quality (12%), and other (4%). Out of the overall tap water quality features of taste, appearance, smell, healthiness, safety, and overall quality, tap water taste (40%) ranked as the number one reason for BW drinkers followed by safety (25%), overall quality (5%), and healthiness (25%). TWA drinkers shown as responding "other" specifically cited convenience, availability, or that fact that "everyone else is doing it". Such motivators for using tap water alternatives were typically cited less than 10% of the time.

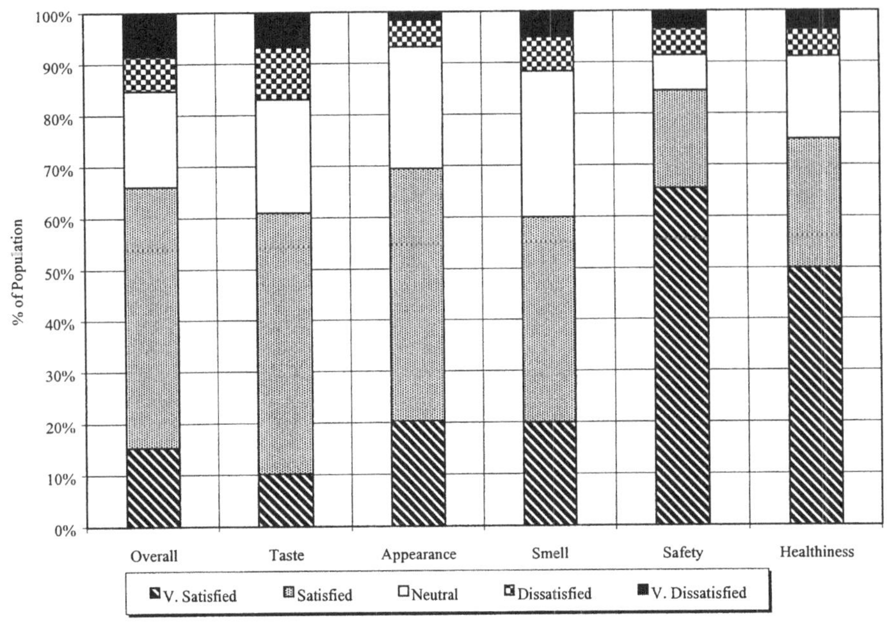

Figure 4.5 TW drinkers' satisfaction levels for various tap water quality issues - Southwest market (Q005, Q010, Q015, Q020, Q025, Q026)

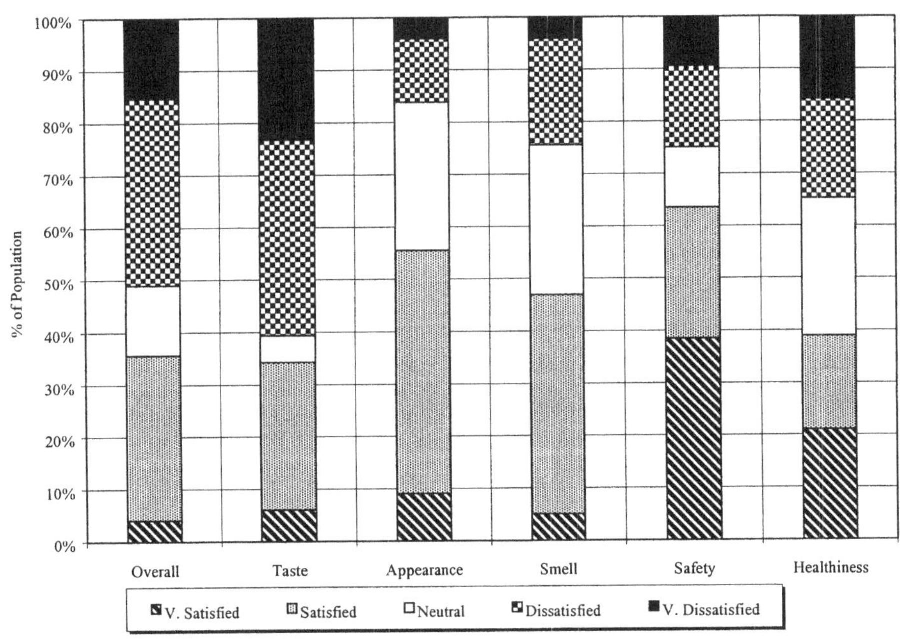

Figure 4.6 TWA drinkers' satisfaction levels for various tap water quality issues - Southwest market (Q005, Q010, Q015, Q020, Q025, Q026)

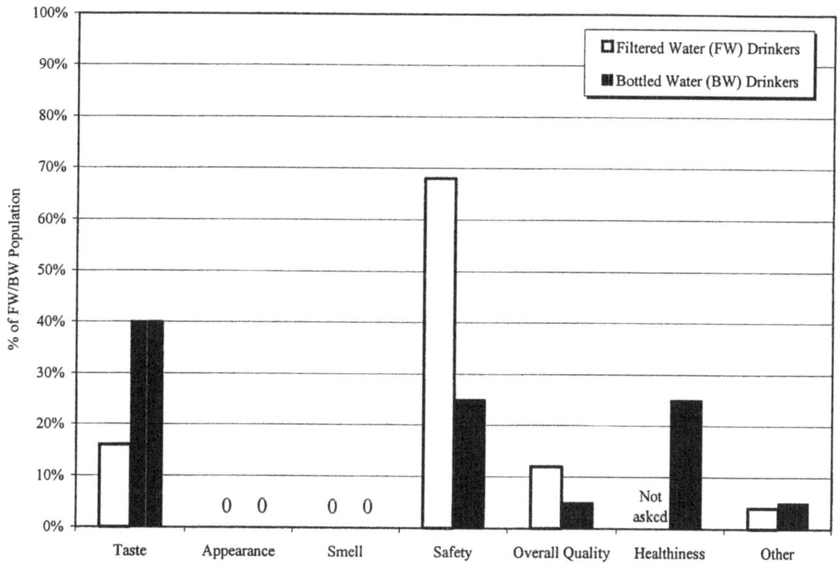

Figure 4.7 Most important reasons for drinking tap water alternatives as cited by BW and FW drinkers - Southwest market (Q091, Q102)

The taste of tap water as received in the household was described by TW and TWA drinkers in Figure 4.8. Half of TW drinkers described the taste of tap water as good versus 29% of TWA drinkers. TWA drinkers qualified tap water taste as being stale/flat (4%), chlorinous (18%), musty/earthy (14%), salty (4%), and metallic (10%).

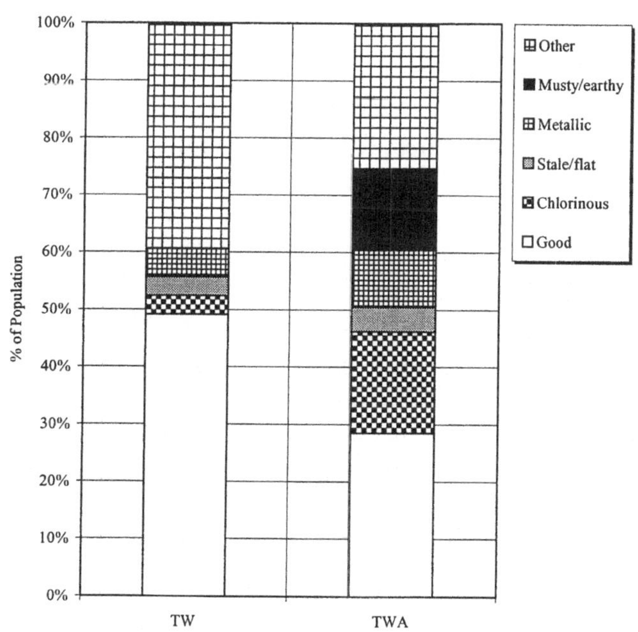

Figure 4.8 Description of tap water taste as perceived by TW and TWA drinkers - Southwest market (Q045)

Consumer Perception of Overall Utility Quality

The Southwest market consumers from both groups described their overall satisfaction with their local utility. Eighty-three percent of TW drinkers were satisfied to very satisfied, as compared to 45% of TWA drinkers. When asked about their satisfaction with resolution of complaints, 29% of TWA drinkers were satisfied (out of 7 total complaints), 29% were neutral, and 43% dissatisfied to very dissatisfied with the complaint resolution of their local provider. Only one tap water user complaint was recorded.

Market West 1

Physical Data

The West 1 market is a private utility owned by a large international water company. The West 1 market facility relies on 89 groundwater wells for 85% of the total water production. The remaining water is supplied by a local river and undergoes a conventional treatment process before distribution into the system.

Water quality for the water is highly variable between well systems but constant for any given well. While some wells supply a water that could be classified as "mineral water" due to the high TDS (>250 mg/L), other wells are quite low in dissolved solids, 36 mg/L. Chlorination is the only treatment applied to the groundwater at the time of the survey. Due to localized complaints of "red water" (i.e., the appearance of solids or cloudiness in the tap water), certain wells have had additional treatment processes put in place.

Consumer Perception of Water Quality

Perceptions of the West 1 market TW drinkers and TWA drinkers of their tap water quality features are illustrated in Figures 4.9 and 4.10. Regarding satisfaction with the overall quality of tap water, 83% of TW drinkers were satisfied to very satisfied, as compared to 64% of TWA drinkers. Safety was rated the highest for both consumer groups (92% for TW drinkers and 80% for TWA drinkers). Fewer satisfied consumers were observed in the taste feature of tap water for the TWA group (49% compared to 84% for TW group). A high number of TWA drinkers were very dissatisfied with taste of tap water (14%).

BW and FW drinkers were asked to cite the most important reason they used tap water alternatives. The results are illustrated in Figure 4.11. Out of the tap water quality features of taste, appearance, smell, safety, and overall quality, tap water taste ranked as the number one reason for FW drinkers (57%) followed by safety (32%), and overall quality (11%). Out of the overall tap water quality features of taste, appearance, smell, healthiness, safety, and overall quality, tap water safety (38%) ranked as the number one reason for BW drinkers followed taste (29%), healthiness (13%), and overall quality (8%).

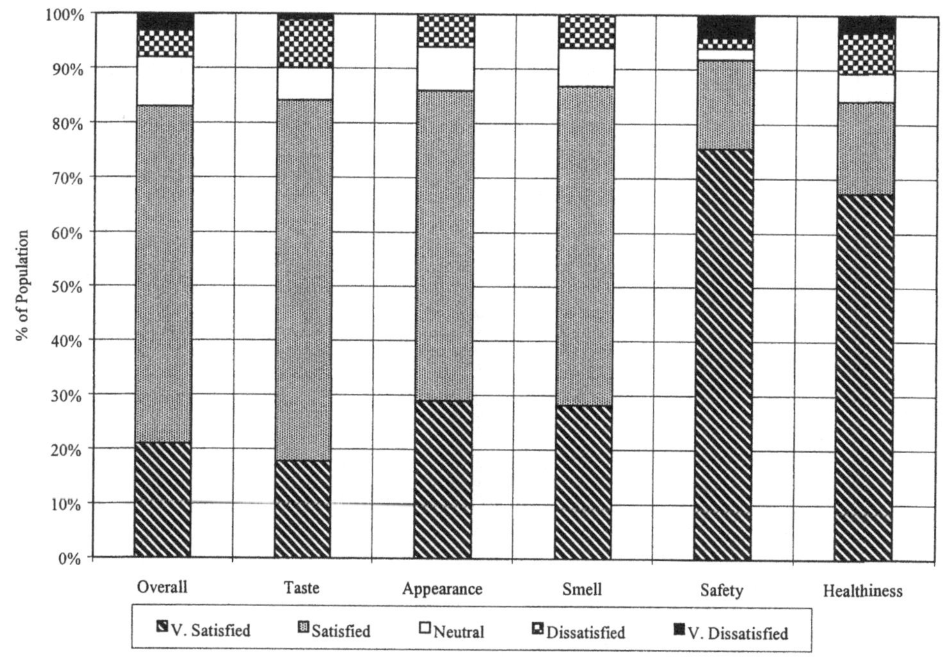

Figure 4.9 TW drinkers' satisfaction levels for various tap water quality issues - West 1 market (Q005, Q010, Q015, Q020, Q025, Q026)

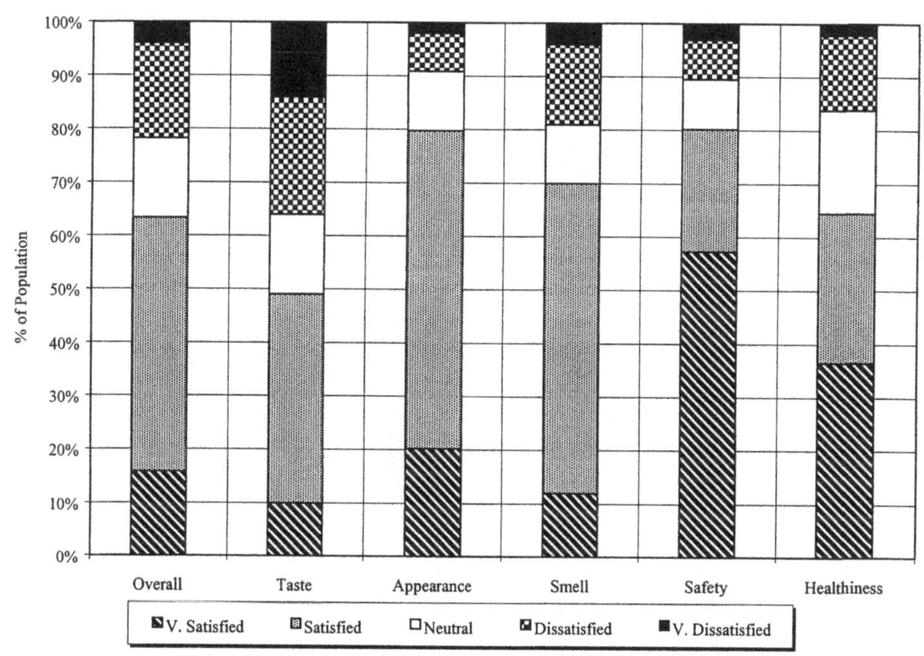

Figure 4.10 TWA drinkers' satisfaction levels for various tap water quality issues - West 1 market (Q005, Q010, Q015, Q020, Q025, Q026)

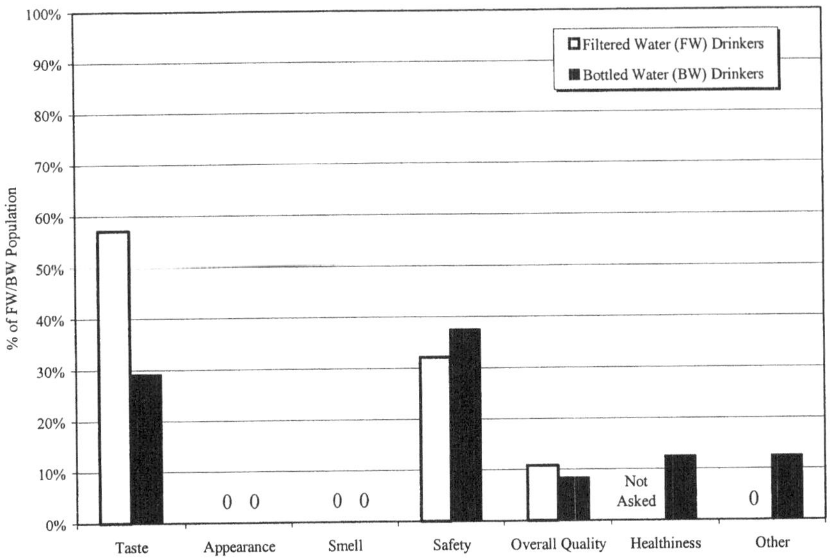

Figure 4.11 Most important reasons for drinking tap water alternatives as cited by BW and FW drinkers - West 1 market (Q091, Q102)

The taste of tap water as received in the household was described by TW and TWA drinkers in Figure 4.12. Fifty-seven percent of TW drinkers described the taste of tap water as good versus 35% of TWA drinkers. TWA drinkers qualified tap water taste as being stale/flat (2%), chlorinous (26%), musty/earthy (3%), and metallic (3%).

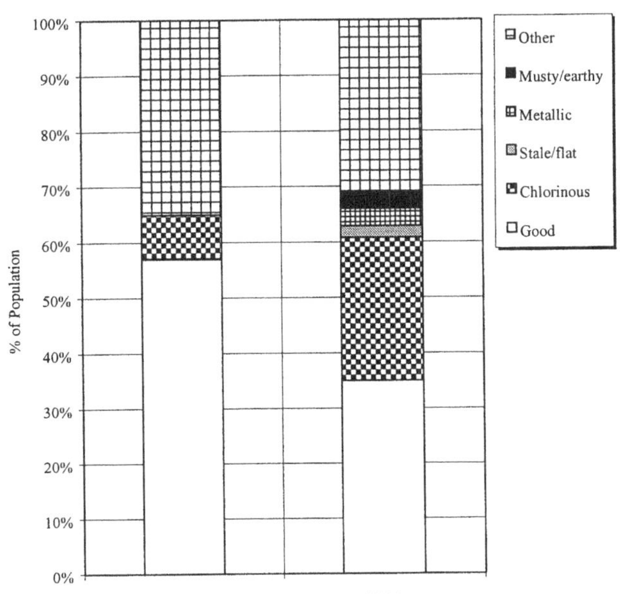

Figure 4.12 Description of tap water taste as perceived by TW and TWA drinkers - West 1 market (Q045)

Consumer Perception of Overall Utility Quality

The West 1 market consumers from both groups described their overall satisfaction with their local utility. Eighty-five percent of TW drinkers were satisfied to very satisfied as compared to 66% of TWA drinkers. When asked about their satisfaction with resolution of complaints, TWA drinkers and TW drinkers were equally satisfied (67%) with the complaint resolution of their local provider (15 complaints were recorded for both TW and TWA drinkers respectively).

Market West 2

Physical Data

Located east of the Continental Divide, the West 2 market is supplied by mountain snow runoff traveling as far away as 200 miles. While the source water has relatively low exposure to anthropogenic contamination, the utility receives about 800 complaints per year. Most complaints focus on appearance related issues (i.e., color or turbidity), while about 10% are taste and odor related (Figure 4.13).

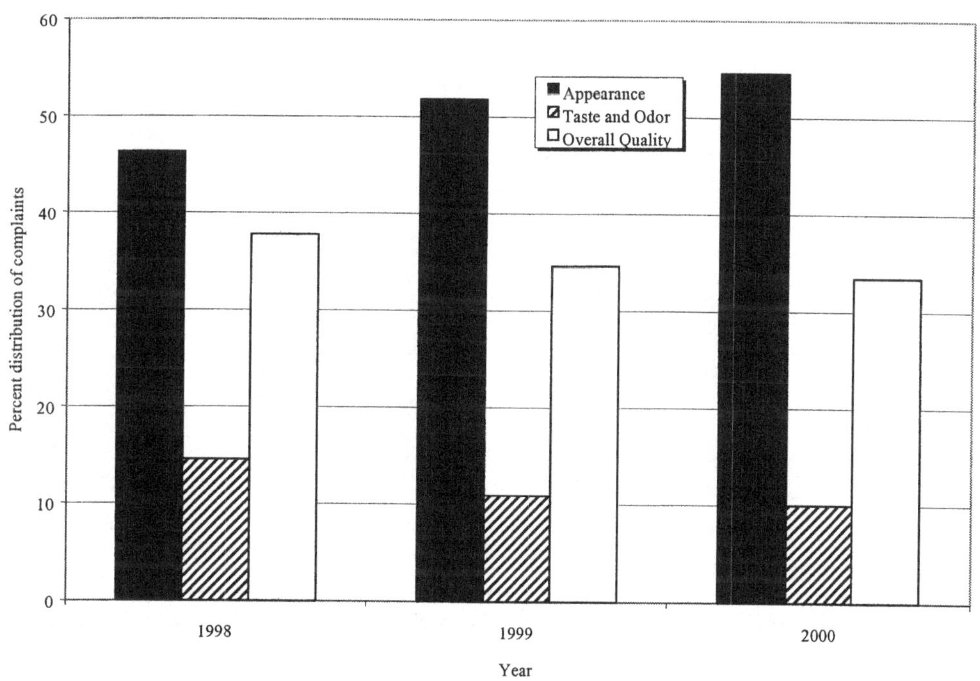

Figure 4.13 Distribution of water related complaints received by West 2 market

Water quality for the West 2 utility consists of a low alkalinity/hardness treated water with low dissolved salts. With cold mountain water as a source, water temperatures are 9°C on average, reaching 3°C in the winter months. With these water qualities, it is most probable that complaints arise from localized problems (e.g., dead-end mains, high chlorine levels for the first few consumers, etc.) rather than poor source water quality.

Consumer Perception of Water Quality

Perceptions of the West 2 market TW drinkers and TWA drinkers of their tap water quality features are illustrated in Figures 4.14 and 4.15. Regarding satisfaction with the overall quality of tap water, 86% of TW drinkers were satisfied to very satisfied, as compared to 67% of TWA drinkers. Similar trends in perceptions were observed in both groups of consumers (i.e., TW drinkers and TWA drinkers). High proportions of satisfied consumers were observed in the smell and safety features of tap water (94–96% for TW drinkers and 78–80% for TWA drinkers).

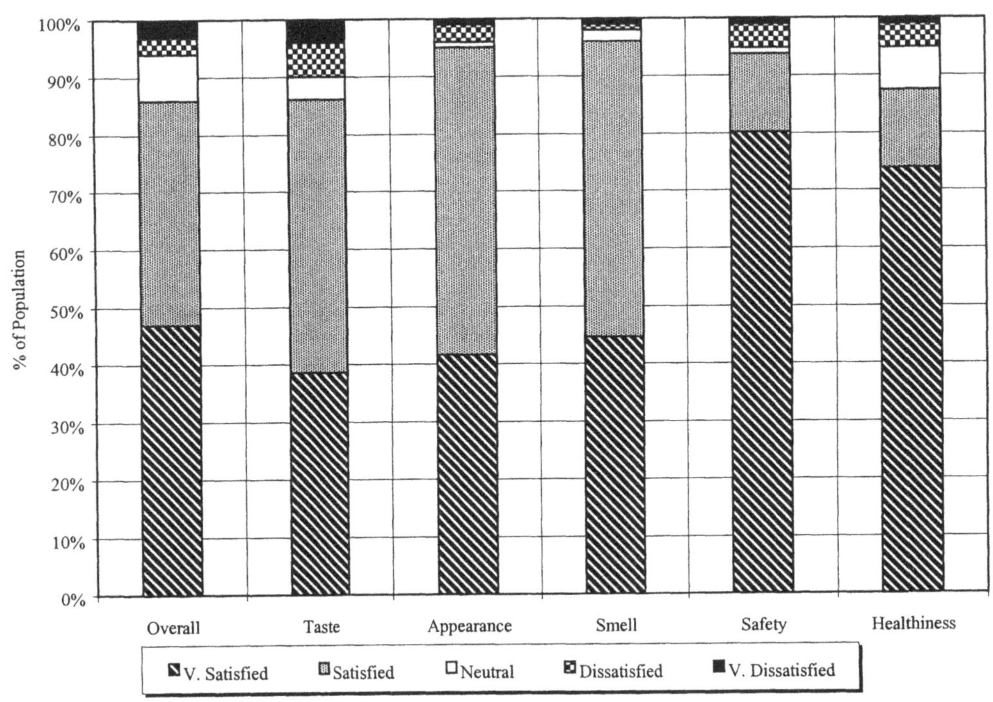

Figure 4.14 TW drinkers' satisfaction levels for various tap water quality issues - West 2 market (Q005, Q010, Q015, Q020, Q025, Q026)

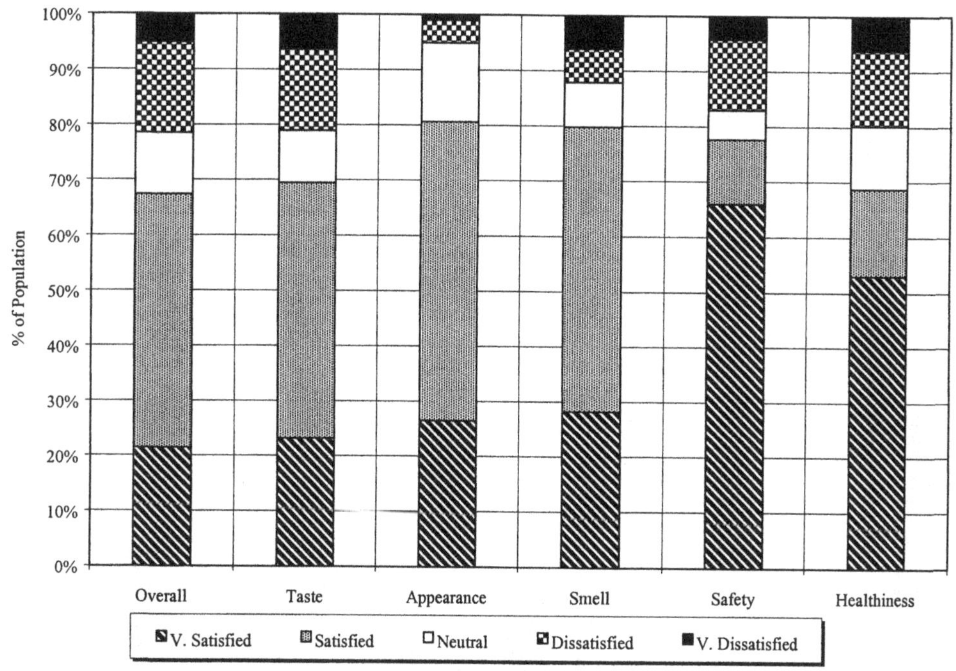

Figure 4.15 TWA drinkers' satisfaction levels for various tap water quality issues - West 2 market (Q005, Q010, Q015, Q020, Q025, Q026)

Fewer satisfied consumers were observed in the overall quality, taste, and healthiness features of tap water (86–88% for TW drinkers and 67–69% for TWA drinkers). Satisfaction with appearance was also high, despite the color problems indicated by consumer complaint data. Overall, there were very few consumers who were very dissatisfied with any of the six tap water quality features (1–4% for TW drinkers as compared to 1–6% for TWA drinkers).

BW and FW drinkers were asked to cite the most important reason they used tap water alternatives. The results are illustrated in Figure 4.16. Out of the tap water quality features of taste, appearance, smell, safety, and overall quality, tap water safety ranked as the number one reason for FW drinkers (70%) followed by taste (20%), and overall quality (10%). Out of the overall tap water quality features of taste, appearance, smell, healthiness, safety, and overall quality, tap water taste (42%) ranked as the number one reason for BW drinkers followed by safety (23%), overall quality (8%), healthiness (8%), and other reasons (19%). These observations are made despite the fact that 69% of BW and FW drinkers were satisfied to very satisfied with the taste of their tap water and 78% were satisfied to very satisfied with the safety of their tap water.

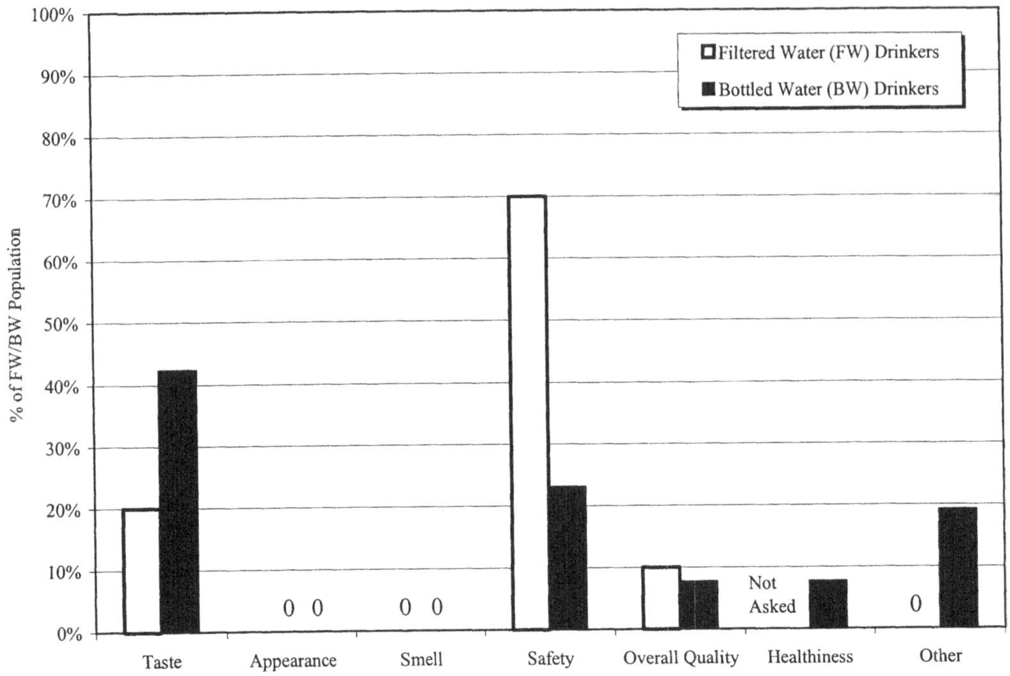

Figure 4.16 Most important reasons for drinking tap water alternatives as cited by BW and FW drinkers - West 2 market (Q091, Q102)

The taste of tap water as received in the household was described by TW and TWA drinkers in Figure 4.17. Seventy percent of TW drinkers described the taste of tap water as good (i.e., fresh or no perceived taste) versus 58% of TWA drinkers. Ten percent of TWA drinkers perceived tap water as having chlorinous flavors versus 6% of TW drinkers. A low percentage of both groups described tap water taste as metallic (1–3%). Seven percent of TWA drinkers described tap water taste as being musty/earthy, as compared to 1% of TW drinkers.

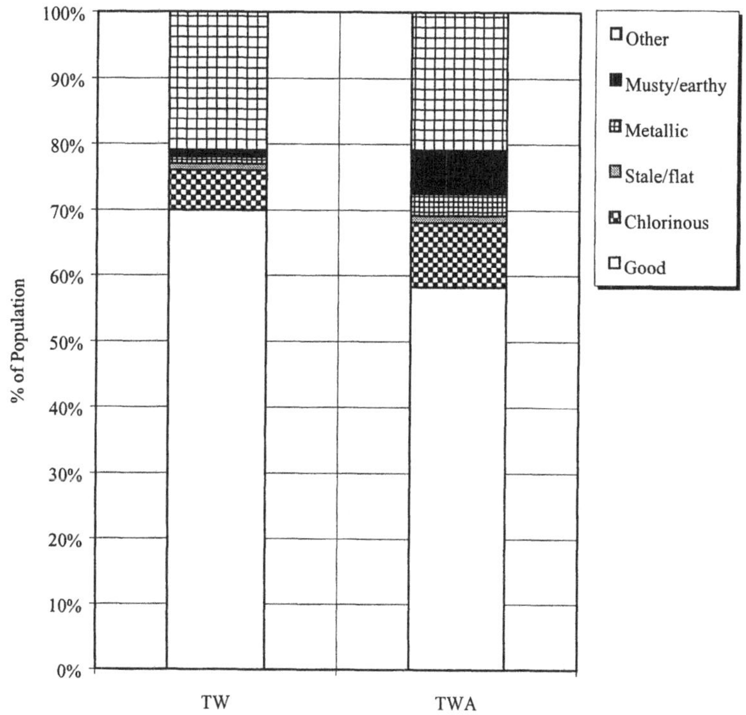

Figure 4.17 Description of tap water taste as perceived by TW and TWA drinkers - West 2 market (Q045)

Consumer Perception of Overall Utility Quality

The West 2 market consumers from both groups described their overall satisfaction with the utility. Ninety percent of TW drinkers were satisfied to very satisfied, as compared to 72% of TWA drinkers. When asked about their satisfaction with resolution of complaints, 100% of TW drinkers were very satisfied to satisfied (6 complaints). The majority (88%) of TWA drinkers opinions ranged from satisfied to very satisfied, while the remainder was dissatisfied with handling of complaints by their local provider (9 complaints total, no neutral responses).

Market Midwest 1

Physical Data

Utility ownership is usually well defined as either public or private. The Midwest 1 market is an unusual "quasi-municipal" system. While the service population is represented by a public board, the water utility operates predominately from revenues generated from the sale of water; state or local funding is minimal.

Water quality for the Midwest 1 market is typical of a Midwestern utility using a surface water. Raw water hardness levels range between 200–300 mg/L as $CaCO_3$ for the Missouri and Kansas rivers but is reduced by about 50% during the softening stage of treatment. Besides high levels of hardness, turbidity can spike up to 2000 NTU and temperature can swing 24°C between seasons.

Consumer Perception of Water Quality

Perceptions of the Midwest 1 market TW drinkers and TWA drinkers of their tap water quality features are illustrated in Figures 4.18 and 4.19. Regarding satisfaction with the overall quality of tap water, 80% of TW drinkers were satisfied to very satisfied, as compared to 52% of TWA drinkers. Similar trends in perceptions were observed in both groups of consumers (i.e., TW drinkers and TWA drinkers). High proportions of satisfied consumers were observed in appearance, smell, and safety features (80 to 90% for TW drinkers and 70 to 77% for TWA drinkers). Fewer satisfied consumers were observed in the taste feature of tap water for TWA drinkers (55%). Overall, there were very few consumers who were very dissatisfied with any of the six tap water quality features (0–1% for TW drinkers as compared to 1–5% for TWA drinkers).

BW and FW drinkers were asked to cite the most important reasons attributed for using tap water alternatives. The results are illustrated in Figure 4.20. Out of the overall tap water quality features of taste, appearance, smell, safety, and overall quality, tap water safety ranked as the number one reason for FW drinkers (55%) followed by taste (40%), and overall quality (5%). Out of the overall tap water quality features of taste, appearance, smell, healthiness, safety, and overall quality, tap water safety (38%) ranked as the number one reason for BW drinkers followed by taste (21%), healthiness (25%), overall quality (13%), and appearance (4%). These observations are made despite the fact that 72 % of TWA drinkers were satisfied to very satisfied with the safety of tap water.

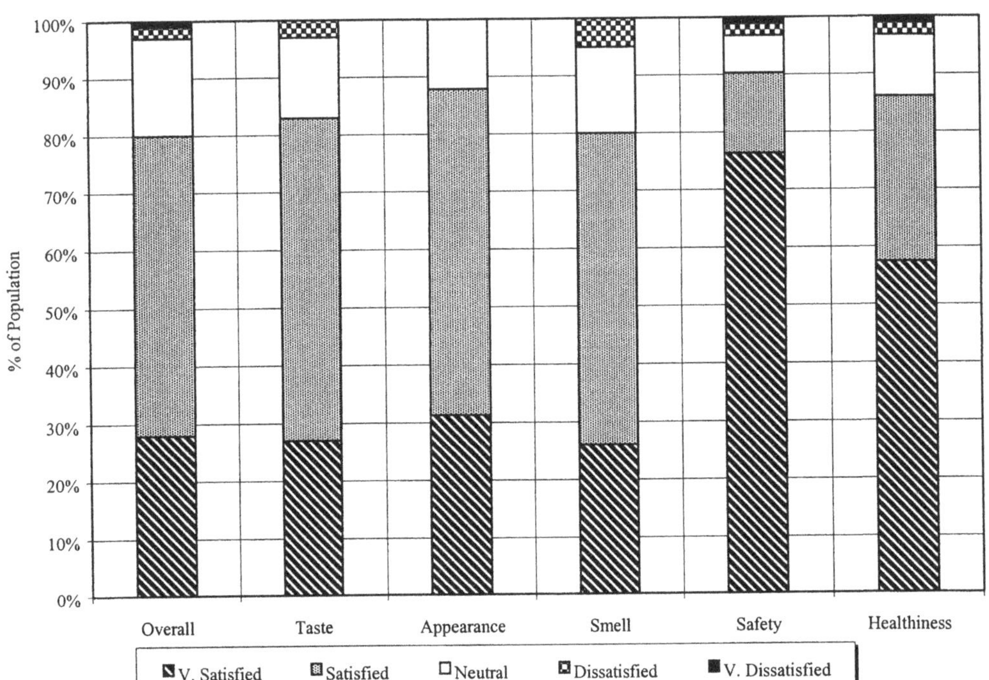

Figure 4.18 TW drinkers' satisfaction levels for various tap water quality issues – Midwest 1 market (Q005, Q010, Q015, Q020, Q025, Q026)

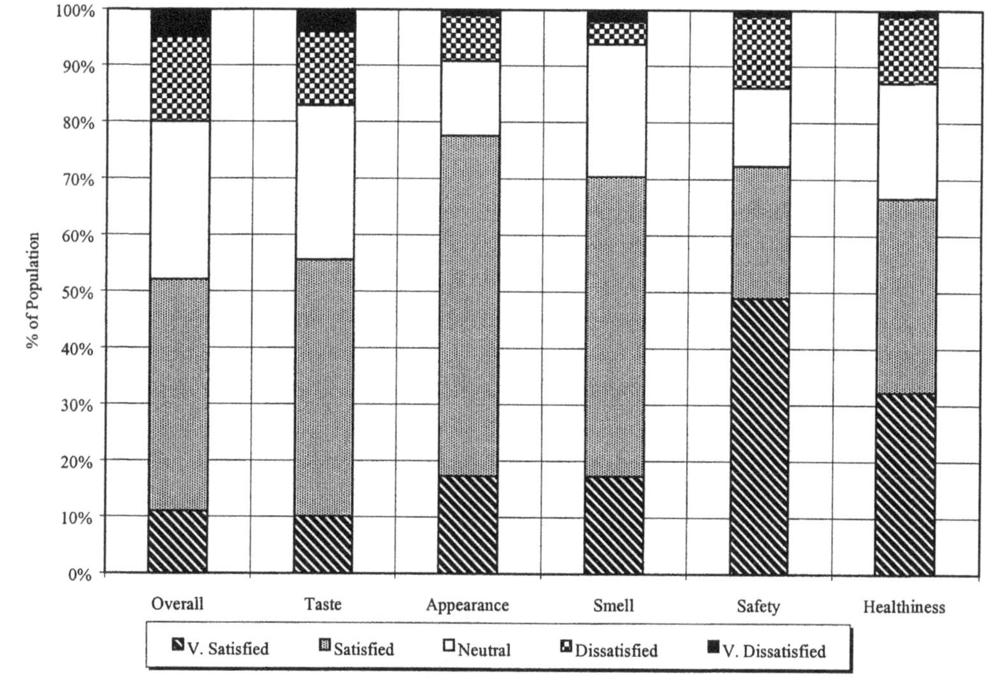

Figure 4.19 TWA drinkers' satisfaction levels for various tap water quality issues - Midwest 1 market (Q005, Q010, Q015, Q020, Q025, Q026)

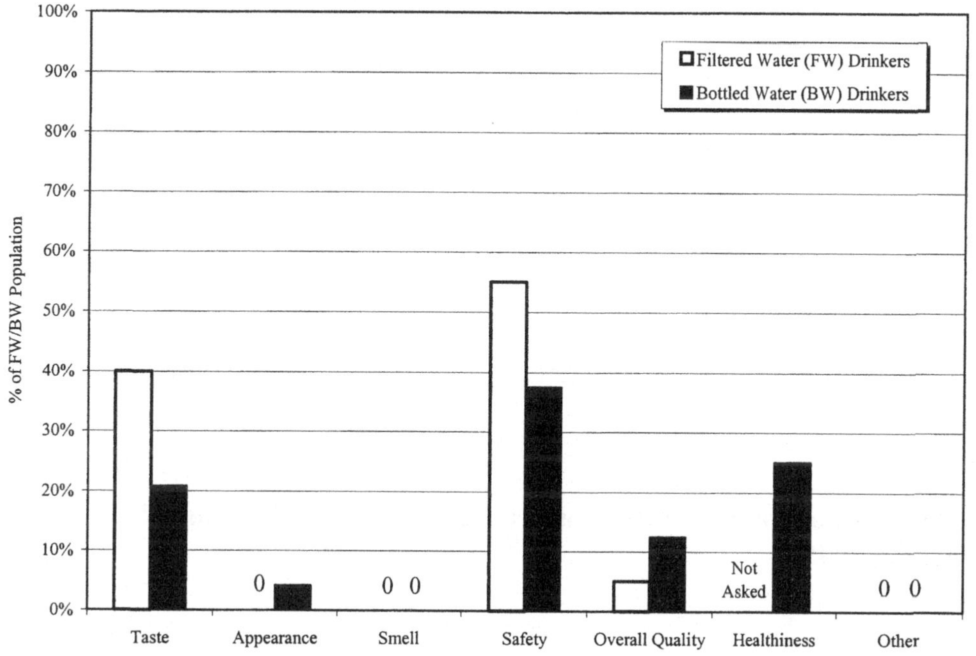

Figure 4.20 Most important reasons for drinking tap water alternatives as cited by BW and FW drinkers - Midwest 1 market (Q091, Q102)

The taste of tap water as received in the household was described by TW and TWA drinkers in Figure 4.21. Eighty-two percent of TW drinkers described the taste of tap water as good (i.e., fresh or no perceived taste) versus 50% of TWA drinkers. Fifteen percent of TWA drinkers perceived tap water as having chlorinous flavors versus 3% of TW drinkers. Twenty-one percent of TWA drinkers described tap water taste as being musty/earthy and stale/flat, while these categories did not apply to TW drinkers.

Consumer Perception of Overall Utility Quality

The Midwest 1 market consumers from both groups described their overall satisfaction with their local utility. Eighty-six percent of TW drinkers were satisfied to very satisfied with Midwest 1 as compared to 70% of TWA drinkers. When asked about their satisfaction with resolution of complaints, lower proportions in both groups were satisfied (50% TW drinkers and 20% TWA drinkers). Sixty percent of TWA drinkers were dissatisfied to very dissatisfied with the complaint resolution of their local provider. A total of six complaints were recorded for each user group.

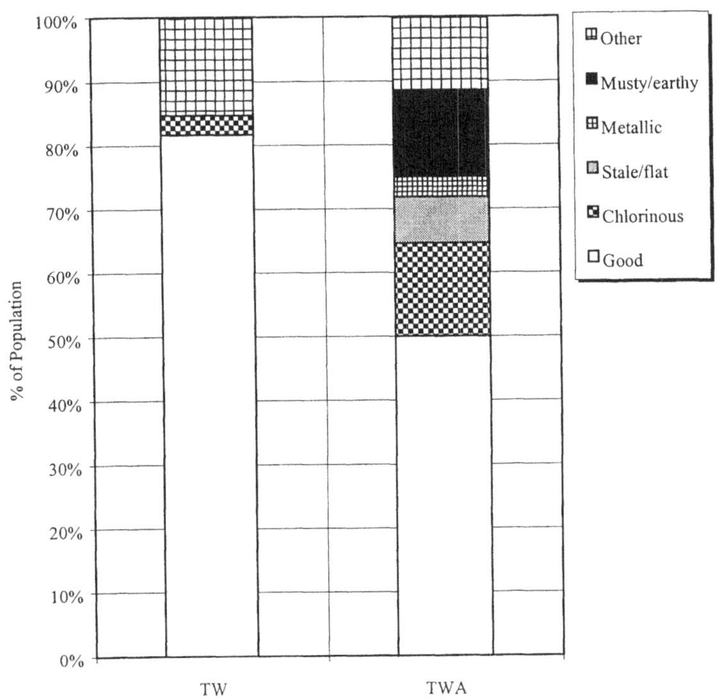

Figure 4.21 Description of tap water taste as perceived by TW and TWA drinkers - Midwest 1 market (Q045)

Market Midwest 2

Physical Data

Situated in the suburbs of Chicago, this 13-mgd facility provides service to a local population of 27,000. Source water from Lake Michigan is treated by conventional processes which have had difficulties removing taste and odor compounds. However, the establishment of zebra mussels in the Great Lakes has been thought to reduce T&O compounds. By filtering a quart of water per day, sunlight now reaches depths that were hidden before and as a result new plant species have taken over. Geosmin/MIB- producing algae are no longer the primary plant species. Most of the consumer complaints in the Midwest 2 market focus on T&O problems and reach a peak during mid-late summer (Figure 4.22). Color problems were predominantly related to service personnel flushing nearby lines which resulted in nearly residents temporarily receiving reddish-yellow colored water.

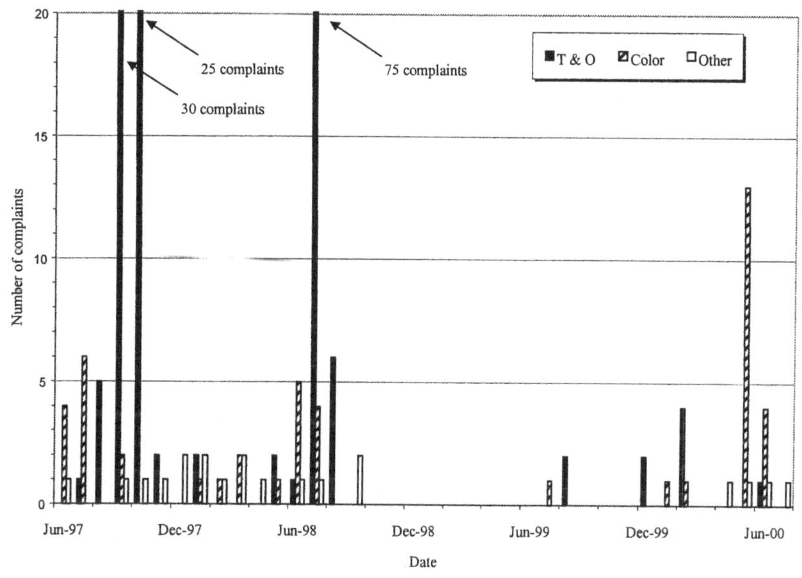

* "T&O" includes earthy, musty; "Color" includes brown, yellow water; "Other" includes all other consumer questions/comments.

Figure 4.22 Variation in frequency and nature of consumer complaints - Midwest 2 market

Water quality for this market remains consistent, except for summer T&O episodes. Average temperature and hardness are moderate enough, at 13°C and 136 mg/L as $CaCO_3$, to avoid any consumer complaints on a year round basis.

Consumer Perception of Water Quality

Perceptions of the Midwest 2 market TW drinkers and TWA drinkers of their tap water quality features are illustrated in Figures 4.23 and 4.24. Regarding satisfaction with the overall quality of tap water, 90% of TW drinkers were satisfied to very satisfied, as compared to 62% of TWA drinkers. A high level of satisfaction was related to safety of tap water (96%) for TW drinkers, as compared to 78% with TWA drinkers. Fewer satisfied consumers were observed in the taste feature of tap water for the TWA group (60% as compared to 87% for TW group). Overall, there were very few consumers who were very dissatisfied with any of the six tap water quality features (0–6% for TW drinkers, and 1–8% for TWA drinkers).

BW and FW drinkers were asked to cite the most important reason they used tap water alternatives. The results are illustrated in Figure 4.25. Out of the overall tap water quality features of taste, appearance, smell, safety, and overall quality, tap water safety ranked as the number one reason for FW drinkers (50%) followed by taste (33%), and overall quality (11%). Out of the overall tap water quality features of taste, appearance, smell, healthiness, safety, and overall quality, tap water taste (31%) ranked as the number one reason for BW drinkers followed by healthiness (28%), safety (22%), overall quality (14%), and appearance (3%).

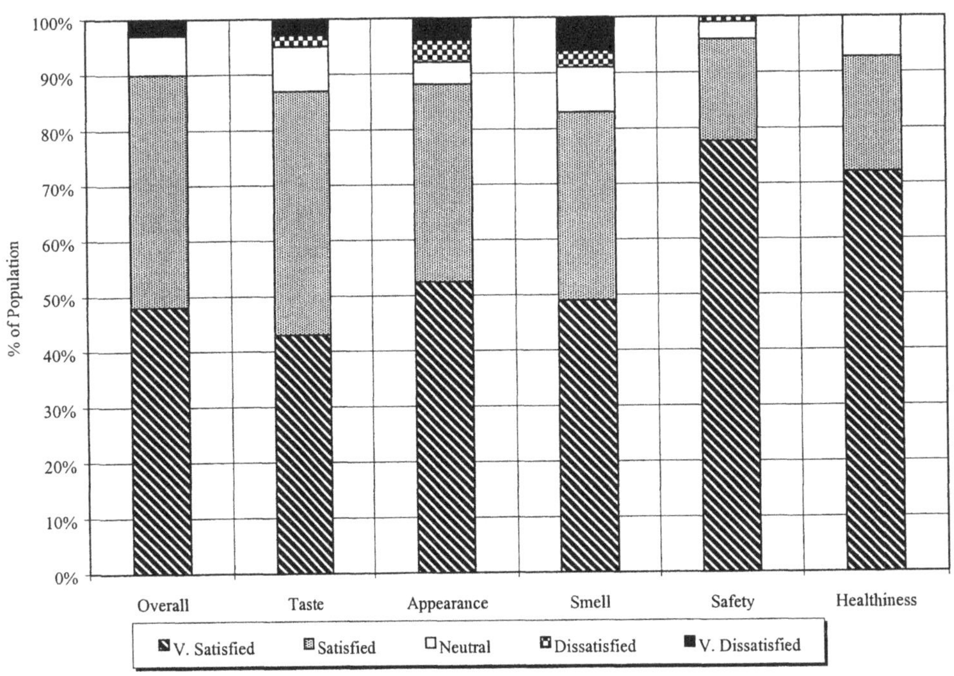

Figure 4.23 TW drinkers' satisfaction levels for various tap water quality issues - Midwest 2 market (Q005, Q010, Q015, Q020, Q025, Q026)

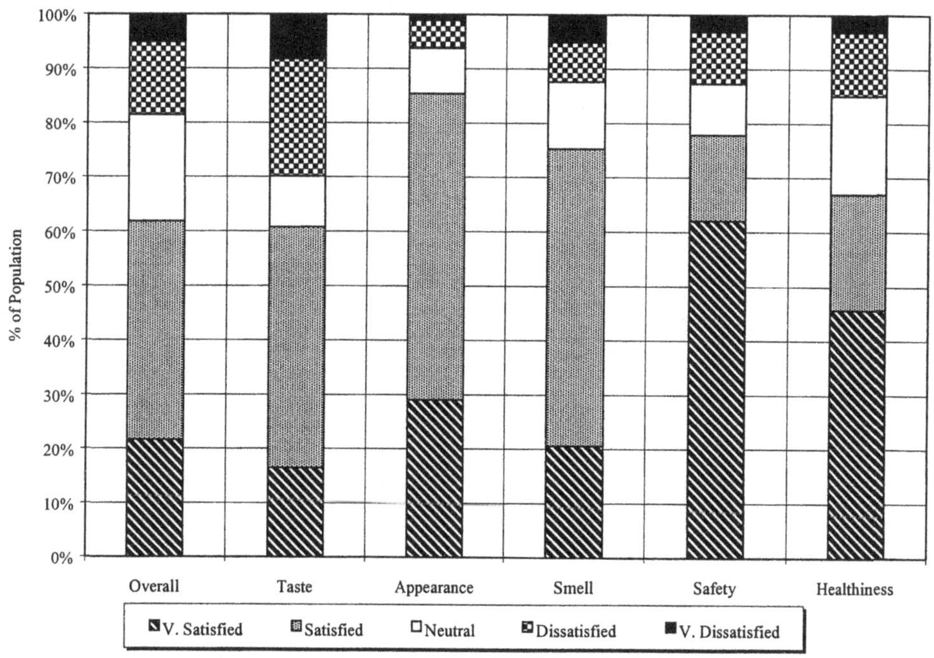

Figure 4.24 TWA drinkers' satisfaction levels for various tap water quality issues - Midwest 2 market (Q005, Q010, Q015, Q020, Q025, Q026)

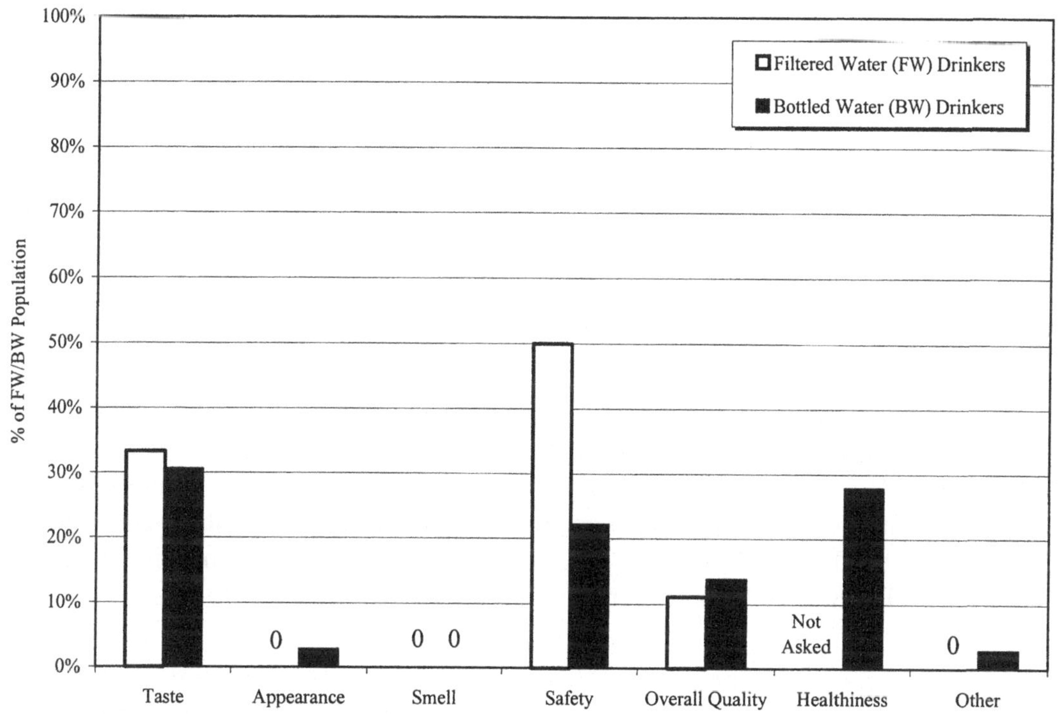

Figure 4.25 Most important reasons for drinking tap water alternatives as cited by BW and FW drinkers - Midwest 2 market (Q091, Q102)

The taste of tap water as received in the household was described by tap water and TWA drinkers in Figure 4.26. Seventy-six percent of TW drinkers described the taste of tap water as good versus 52% of TWA drinkers. TWA drinkers qualified tap water taste as being stale/flat (2%), chlorinous (9%), and metallic (5%).

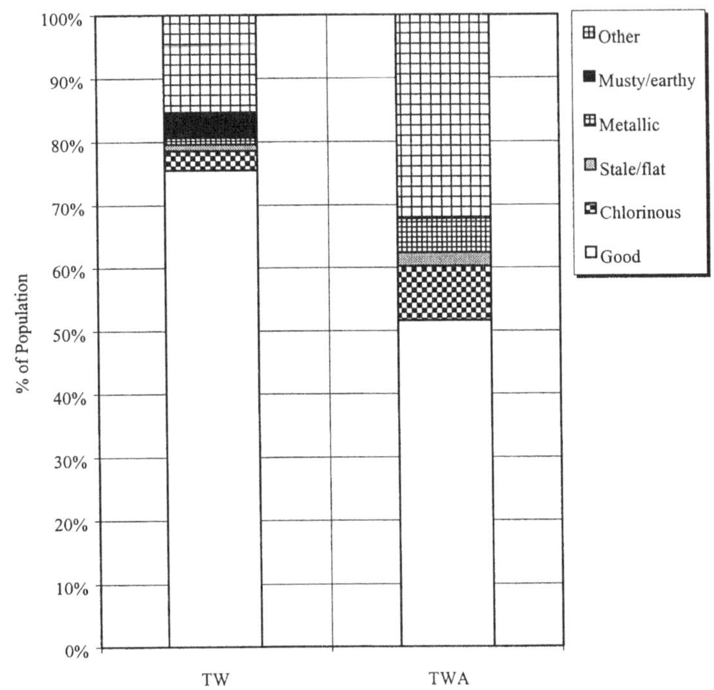

Figure 4.26 Description of tap water taste as perceived by TW and TWA drinkers - Midwest 2 market (Q045)

Consumer Perception of Overall Utility Quality

Midwest 2 consumers from both groups described their overall satisfaction with their local utility. Ninety-five percent of TW drinkers were satisfied to very satisfied, as compared to 76% of TWA drinkers. When asked about their satisfaction with resolution of complaints, TWA drinkers were more satisfied than TW drinkers with the complaint resolution of their local provider (75% for TW drinkers and 84% for TWA drinkers). A total of six and nine complaints were recorded by the TW and TWA consumers respectively.

Market Midwest 3

Physical Data

Water usage can vary between locations based on the local economy. Whereas, the Southeast 2 market supplies over half of their treated water to residents and businesses, the majority of the Midwest 3 market's water, 60%, goes to local industries. Two large chemical manufacturers have sizeable operations in Midwest 3's county. Source water comes from one of the Great Lakes and is piped 65 miles before being treated and available for use by the 75,000 citizens of the county. As a moderate-sized community, Midwest 3 has few water complaints; only 27 for 1998 and 1999. About half of these were related to consumer plumbing. Either old valves or fixtures (e.g., white plastic dip tube in water heaters) caused complaints about appearance of white or black particles.

Water quality for Midwest 3 remains fairly constant with occasional turbidity spikes to 30 NTU following rain events. Source water has a moderate level of hardness and TDS, 101 mg/L as $CaCO_3$ and 111 mg/L, respectively.

Consumer Perception of Water Quality

Perceptions of Midwest 3 TW drinkers and TWA drinkers of their tap water quality features are illustrated in Figures 4.27 and 4.28. Regarding satisfaction with the overall quality of tap water, 95% of TW drinkers were satisfied to very satisfied, as compared to 75% of TWA drinkers. Similar trends in perceptions were observed in both groups of consumers: TW consumers were more satisfied with overall quality, appearance, smell, safety, and healthiness of tap water (89 to 95% for TW drinkers and 71 to 85% for TWA drinkers). Fewer satisfied consumers were observed in the taste feature of tap water for TWA drinkers (64%). Overall, there were very few consumers who were very dissatisfied with any of the six tap water quality features (0% for TW drinkers as compared to 1–5% for TWA drinkers).

BW and FW drinkers were asked to cite the most important reason they used tap water alternatives. The results are illustrated in Figure 4.29. Out of the tap water quality features of taste, appearance, smell, safety, and overall quality, tap water safety ranked as the number one reason for FW drinkers (53%) followed by overall quality (29%), and taste (18%). Out of the overall tap water quality features of taste, appearance, smell, healthiness, safety, and overall quality, tap water taste (42%) ranked as the number one reason for BW drinkers followed by safety (25%), healthiness (25%), and smell (8%). These observations are made despite the fact that 85% of TWA drinkers were satisfied to very satisfied with the safety of tap water.

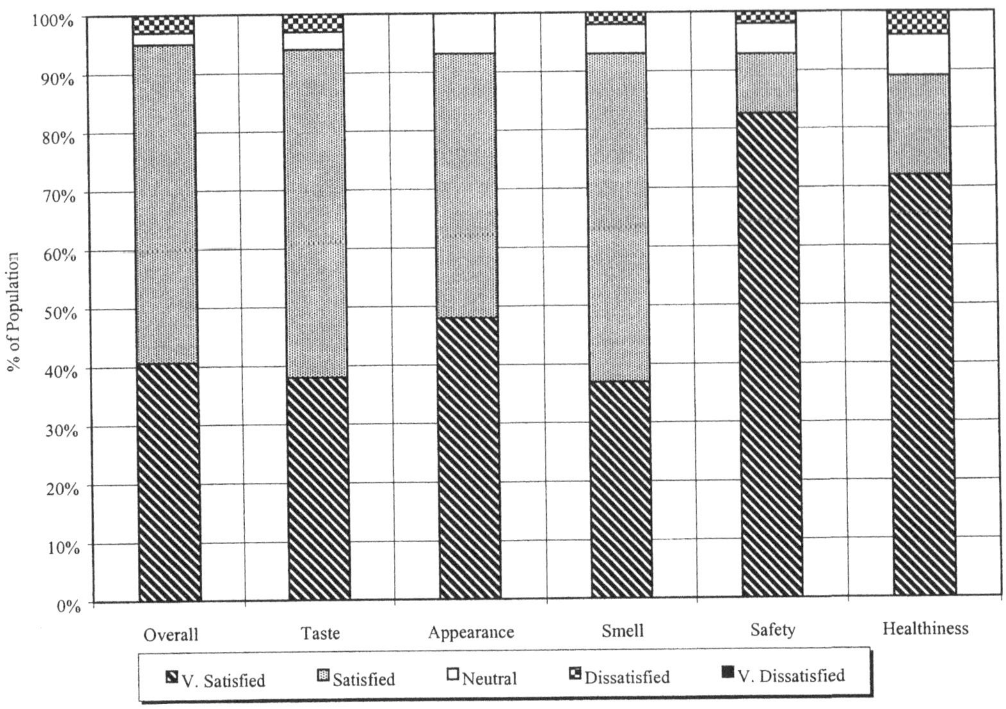

Figure 4.27 TW drinkers' satisfaction levels for various tap water quality issues - Midwest 3 market (Q005, Q010, Q015, Q020, Q025, Q026)

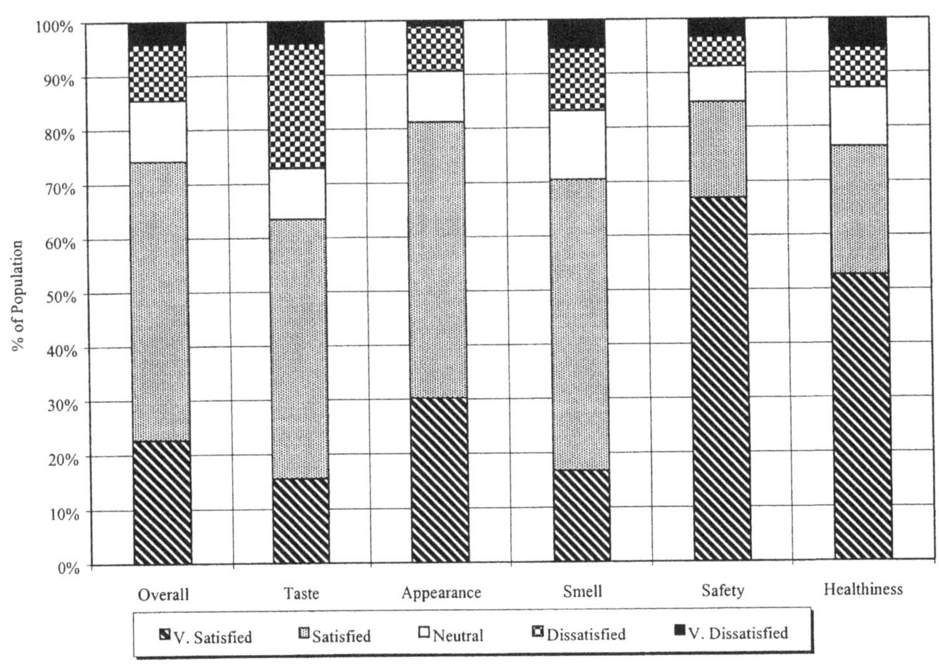

Figure 4.28 TWA drinkers' satisfaction levels for various tap water quality issues - Midwest 3 market (Q005, Q010, Q015, Q020, Q025, Q026)

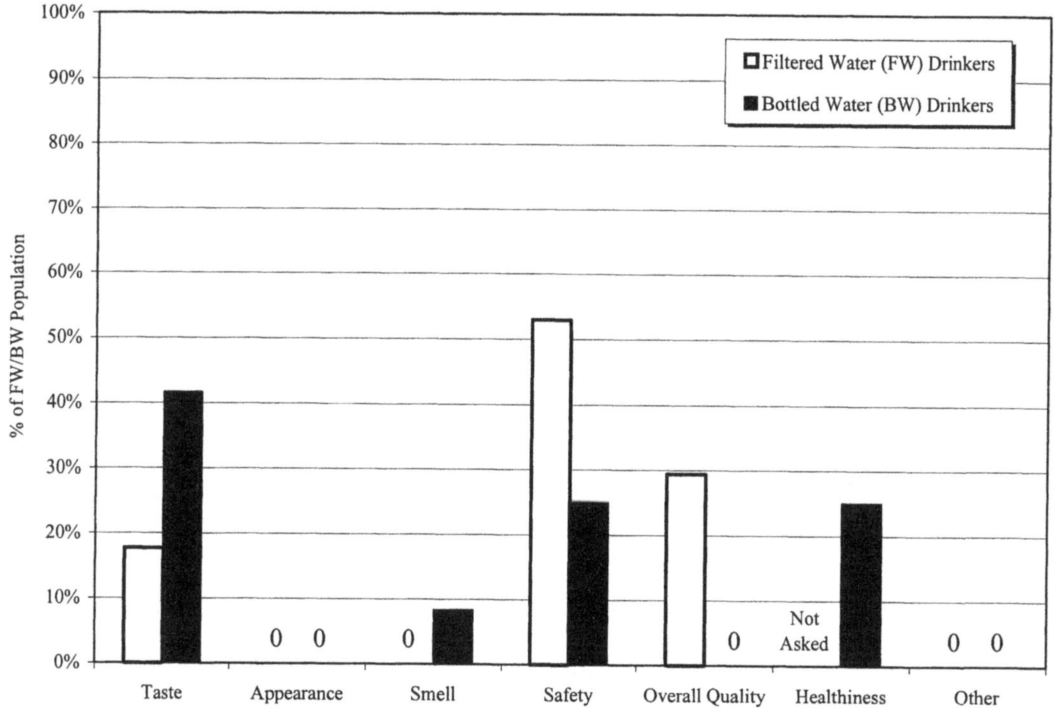

Figure 4.29 Most important reasons for drinking tap water alternatives as cited by BW and FW drinkers - Midwest 3 market (Q091, Q102)

The taste of tap water as received in the household was described by TW and TWA drinkers in Figure 4.30. Sixty-five percent of TW drinkers described the taste of tap water as good versus 44% of TWA drinkers. Thirty-three percent of TWA drinkers perceived tap water as having chlorinous flavors versus 12% of TW drinkers. Other reasons attributed by TWA drinkers to drinking tap water alternatives were musty/earthy and metallic tap water taste (3%).

Consumer Perception of Overall Utility Quality

Midwest 3 consumers from both groups described their overall satisfaction with their local utility. Ninety-one percent of TW drinkers were satisfied to very satisfied as compared to 68% of TWA drinkers. When asked about their satisfaction with resolution of complaints, 86% of TW group were satisfied to very satisfied, while only 57% of TWA drinkers were satisfied to very satisfied (out of 7 complaints reported in each user group).

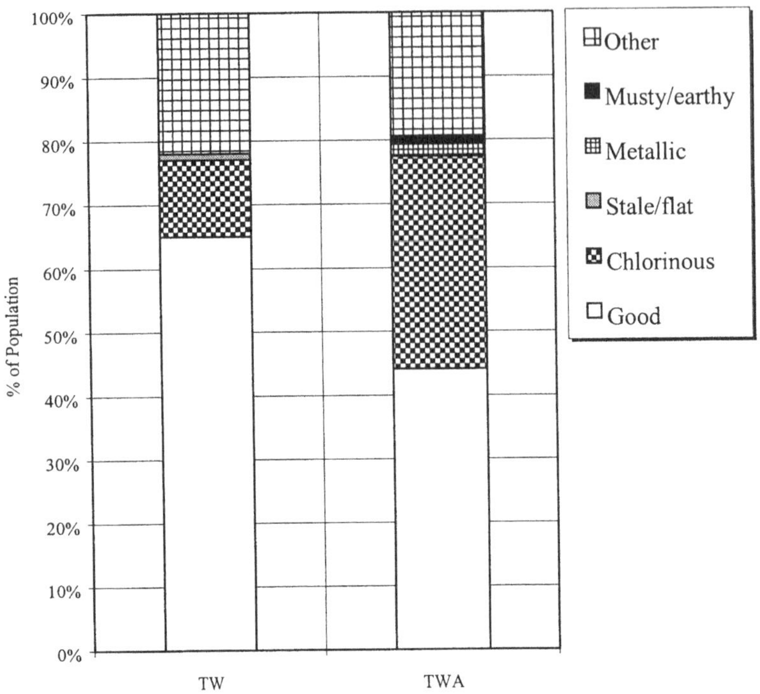

Figure 4.30 Description of tap water taste as perceived by TW and TWA drinkers - Midwest 3 market (Q045)

Market East 1

Physical Data

The East 1 market is one of the largest water systems in the U.S. This utility has experienced a gamut of water issues, ranging from water quantity to water quality issues (Jaffe 1995a and 1995b). The department has weathered these problems due to a well-trained technical and public relations department. For example, in 1982, 41 T&O complaints were received in 3 days all related to a "cucumber" smell. Through careful detective work, the origin of the odor was later identified and the problem mediated. This was the first reported case of a drinking water system experiencing "cucumber" odor (Burlingame *et al.* 1992). Later, in 1985, when problems arose once again, the department was able to pinpoint the specific area causing the problem (Burlingame *et al.* 1986).

Finished water quality is of moderate hardness and TDS, 102 mg/L as $CaCO_3$ and 214 mg/L, respectively. While the average temperature of the finished water is moderate at 15°C, summertime temperatures of the raw surface water can reach 34°C. Treatment modifications (e.g., use of strong oxidant like potassium permanganate) are implemented during these times to control T&O problems that have arisen in the past.

Consumer Perception of Water Quality

Perceptions of the East 1 market TW drinkers and TWA drinkers of their tap water quality features are illustrated in Figures 4.31 and 4.32. Regarding satisfaction with the overall quality of tap water, 75% of TW drinkers were satisfied to very satisfied, as compared to 52% of TWA drinkers. Safety was rated the highest by both consumer groups (89% for TW drinkers and 75% for TWA drinkers). Fewer satisfied consumers were observed in the taste feature of tap water for the TWA group (50% as compared to 78% for TW group). Overall, there were very few consumers who were very dissatisfied with any of the six tap water quality features (0–4% for TW drinkers as compared to 3–9% for TWA drinkers).

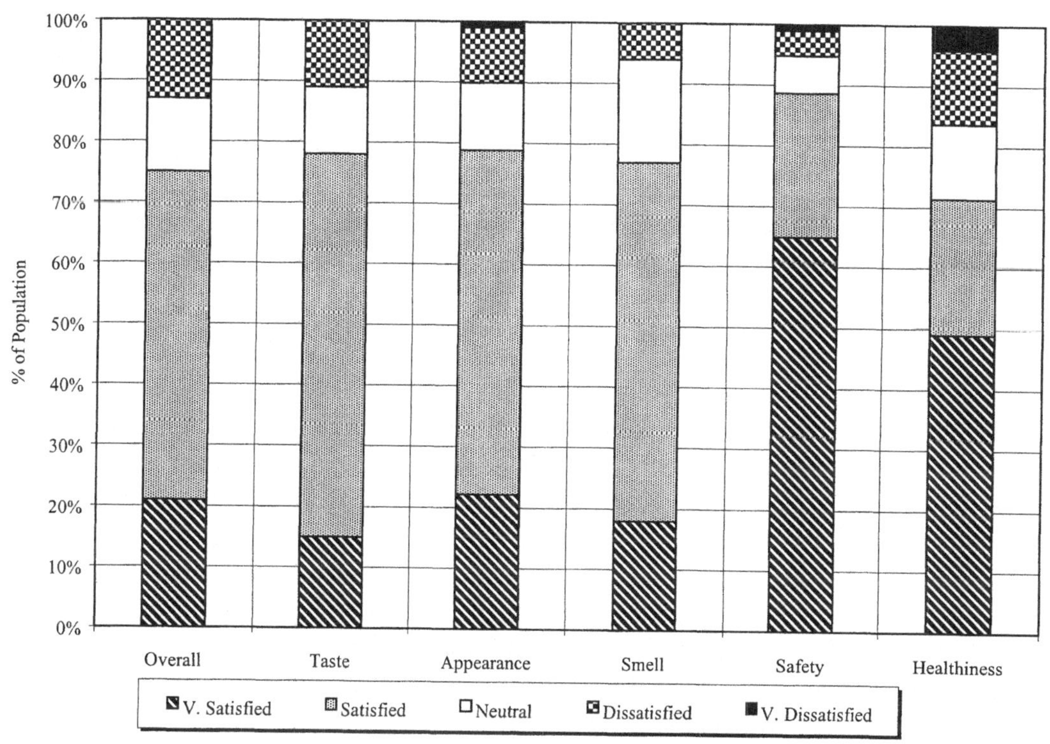

Figure 4.31 TW drinkers' satisfaction levels for various tap water quality issues - East 1 market (Q005, Q010, Q015, Q020, Q025, Q026)

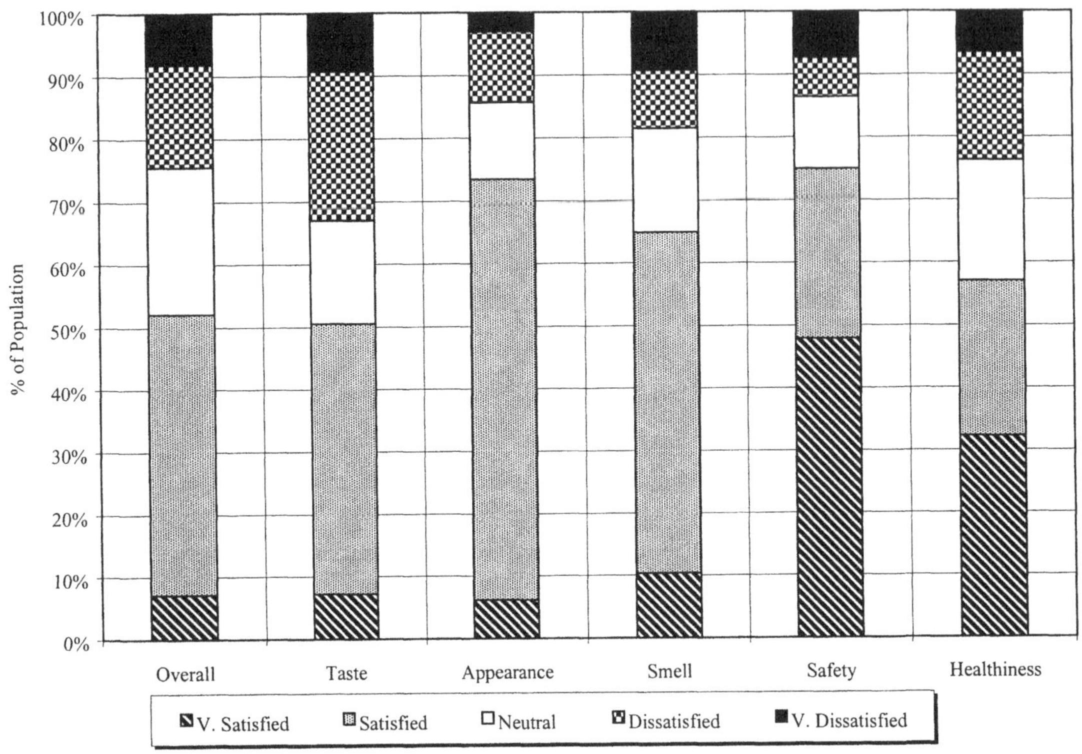

Figure 4.32 TWA drinkers' satisfaction levels for various tap water quality issues - East 1 market (Q005, Q010, Q015, Q020, Q025, Q026)

BW and FW drinkers were asked to cite the most important reason they used tap water alternatives. The results are illustrated in Figure 4.33. Out of the overall tap water quality features of taste, appearance, smell, safety, and overall quality, tap water safety ranked as the number one reason for FW drinkers (74%) followed by overall quality (16%), and taste (11%). Out of the overall tap water quality features of taste, appearance, smell, healthiness, safety, and overall quality, tap water healthiness (35%) ranked as the number one reason for BW drinkers followed by safety (31%), taste (19%), and overall quality (11%). These observations were made despite the fact that 75% of the TWA group perceived tap water as safe to drink.

The taste of tap water as received in the household was described by TW and TWA drinkers in Figure 4.34. Sixty-five percent of TW drinkers described the taste of tap water as good versus 46% of TWA drinkers. Twenty-one percent of TWA drinkers perceived tap water as having chlorinous flavors versus 4% of TW drinkers.

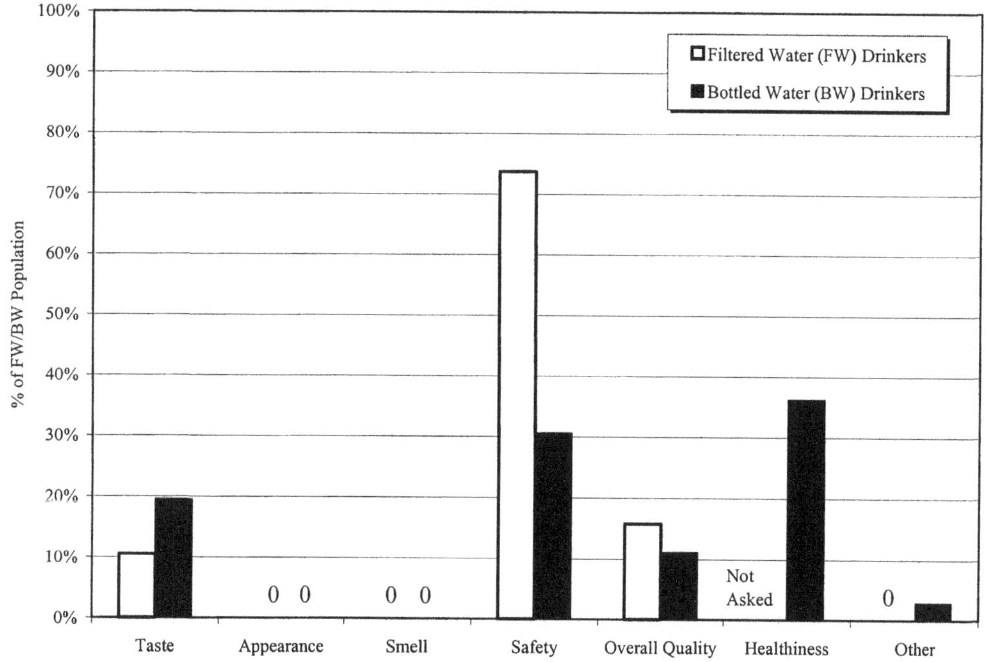

Figure 4.33 Most important reasons for drinking tap water alternatives as cited by BW and FW drinkers - East 1 market (Q091, Q102)

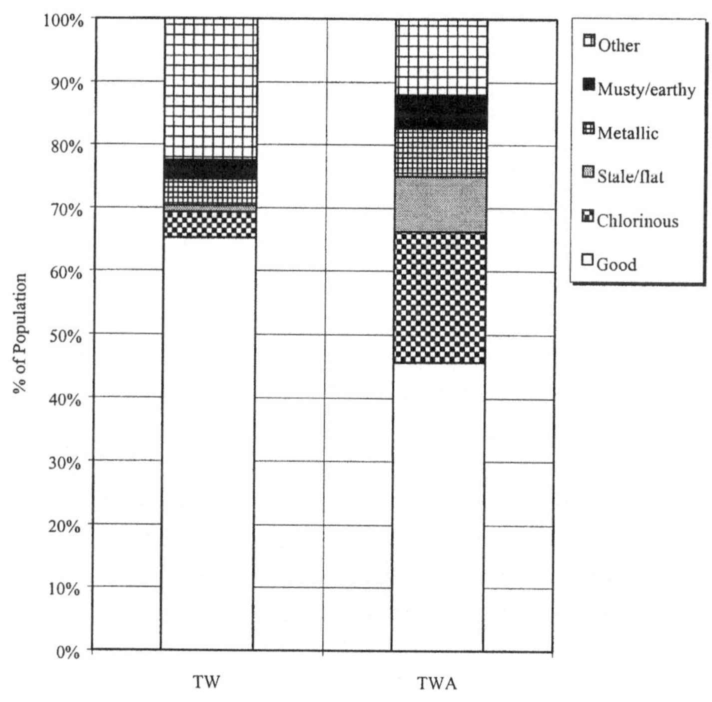

Figure 4.34 Description of tap water taste as perceived by TW and TWA drinkers - East 1 market (Q045)

Consumer Perception of Overall Utility Quality

East 1 consumers from both groups described their overall satisfaction with their local utility. Eighty-two percent of TW drinkers were satisfied to very satisfied with PWD as compared to 58% of TWA drinkers. When asked about their satisfaction with resolution of complaints, only 44% of TW group were satisfied to very satisfied with the complaint resolution of their local provider (out of 14 complaints total). However, 57% of TWA drinkers were satisfied with the complaint resolution of their local provider (out of 7 complaints total).

Market East 2

The East 2 market is a combination of two water utilities in close geographic proximity to one another. Their physical characteristics are defined separately here as market East 2a and East 2b.

Physical Data

Market East 2a. East 2a is a large, private utility serving over 300,000 people. Nine separate reservoirs are used as a supply. In the summer of 2000, a boil order was issued when one of these reservoirs (and the associated distribution system) tested positive for coliform bacteria. A quick response to this problem helped thwart what otherwise would have been a major public incident; consumer complaints actually decreased slightly from 3.3 to 3.2 per day. This veteran public relation/customer satisfaction department was also responsible for quick resolution of consumer complaints in 1999 and 2000. Two out of every three complaints were resolved the same day that the complaint was made. The remaining one-third were complaints that required special site visits to mediate the problem (e.g., pipe flushing or tap water testing) or were out of the hands of the utility (e.g., corroding household plumbing or deteriorating diptubes in the water heater).

The remaining service areas receive primarily surface water supplied by two plants using conventional treatment. Finished water has a low hardness and TDS of 48 mg/L as $CaCO_3$ and 116 mg/L, respectively. Concern has been raised for a very small portion of the service population (less than 9%) that is exposed to the radon found in the groundwater. Measurements up to 3,000 pci/L have been made.

Market East 2b. East 2b is a large public utility using nine separate treatment facilities that draw from local reservoirs. Special watershed protection programs have helped avoid any boil water notices or other contamination issues as experienced by the nearby East 2a market. However, summer heat can affect reservoir water quality resulting in seasonal T&O problems.

Complaints typically peak during the summer months, as shown in Figure 4.35. Moderate levels of hardness and TDS are not high enough to arouse consumer complaint; 70 mg/L as $CaCO_3$ and 138 mg/L, respectively. Some complaints focus on chlorinous or sulfurous smell but most, especially during summer months, focus on earthy/musty odors.

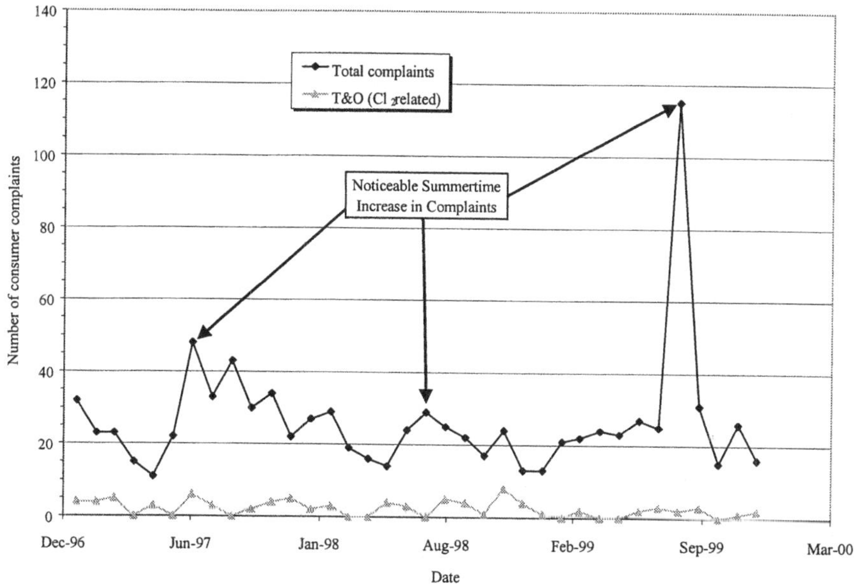

Figure 4.35 Seasonal variation in consumer complaints - East 2b market

Consumer Perception of Water Quality

Perceptions of the East 2 market TW drinkers and TWA drinkers of their tap water quality features are illustrated in Figures 4.36 and 4.37. Regarding satisfaction with the overall quality of tap water, 87% of TW drinkers were satisfied to very satisfied, as compared to 67% of TWA drinkers. Safety was rated the highest by both consumer groups (97% for TW drinkers and 76% for TWA drinkers). Fewer satisfied consumers were observed in the taste feature of tap water (77% for TW drinkers and 55% for TWA drinkers). Overall, there were very few consumers who were very dissatisfied with any of the six tap water quality features (0 to 4% for TW drinkers as compared to 4 to 8% for TWA drinkers).

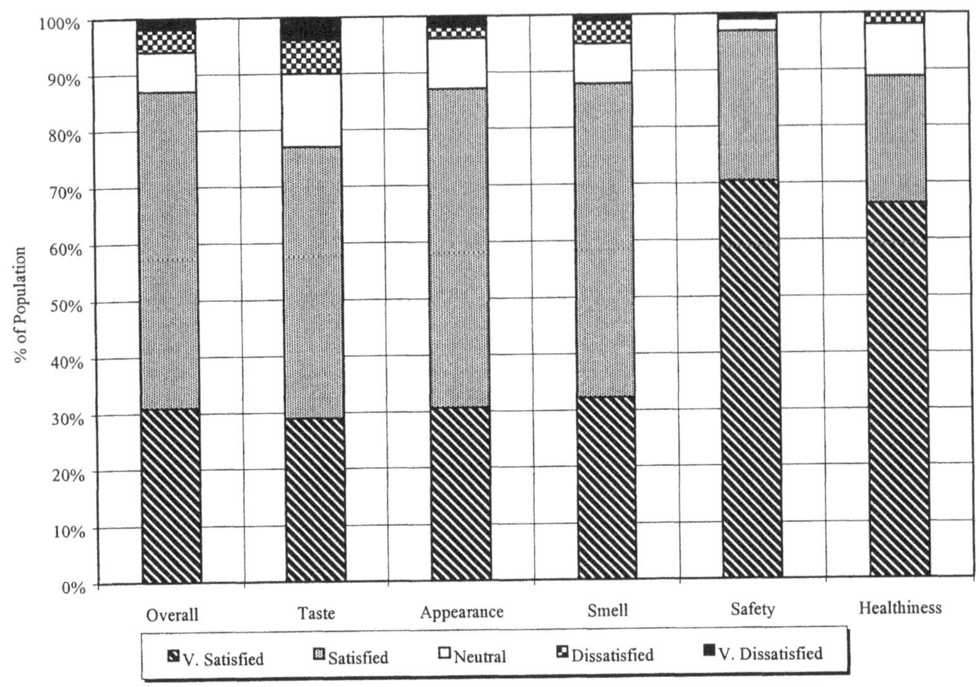

Figure 4.36 TW drinkers' satisfaction levels for various tap water quality issues - East 2 market (Q005, Q010, Q015, Q020, Q025, Q026)

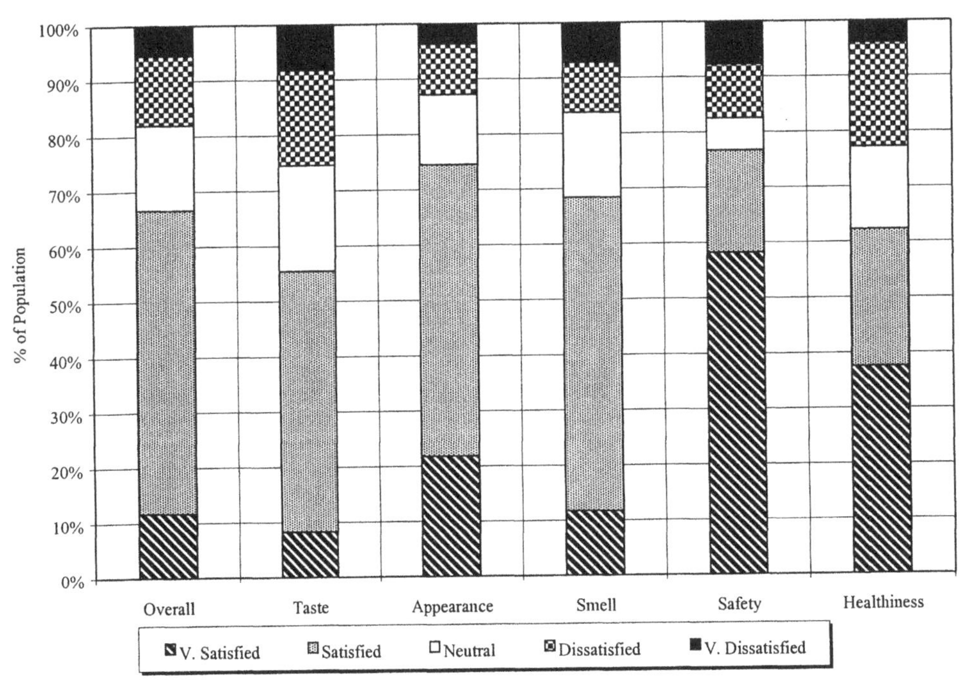

Figure 4.37 TWA drinkers' satisfaction levels for various tap water quality issues - East 2 market (Q005, Q010, Q015, Q020, Q025, Q026)

BW and FW drinkers were asked to cite the most important reason they used tap water alternatives. The results are illustrated in Figure 4.38. Out of the tap water quality features of taste, appearance, smell, safety, and overall quality, tap water safety ranked as the number one reason for FW drinkers (55%) followed by taste (32%). Out of the overall tap water quality features of taste, appearance, smell, healthiness, safety, and overall quality, tap water taste and safety (33% each) ranked as the number one reasons for BW drinkers followed by healthiness (28%). These observations are made despite the fact that 76% of TWA drinkers were satisfied to very satisfied with the safety of tap water.

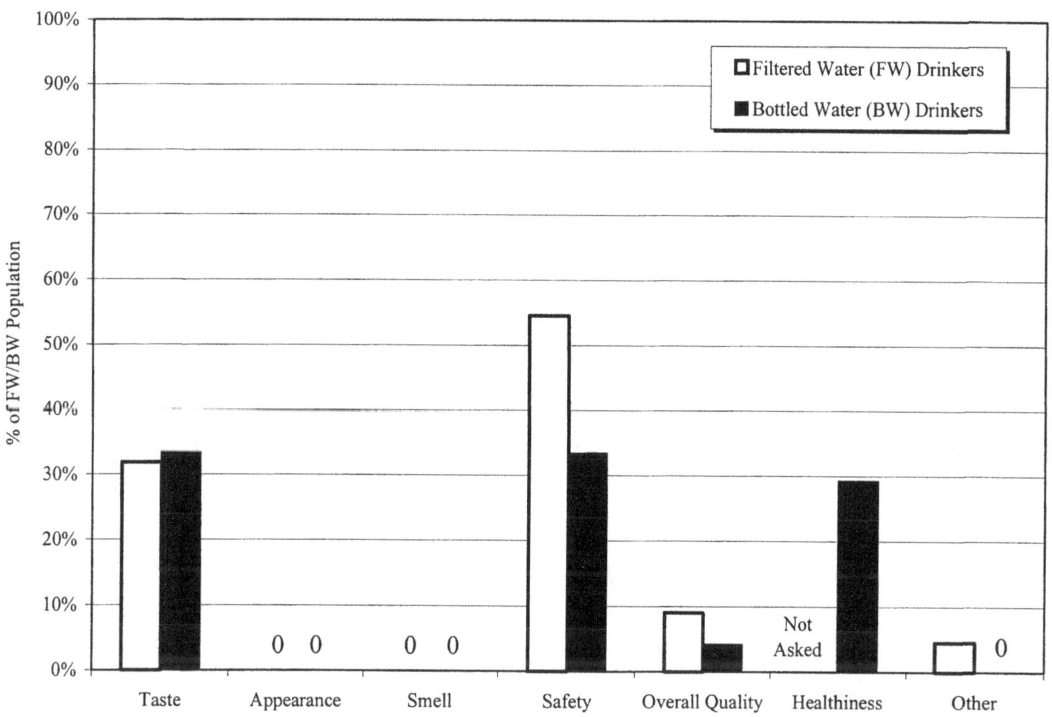

Figure 4.38 Most important reasons for drinking tap water alternatives as cited by BW and FW drinkers - East 2 market (Q091, Q102)

The taste of tap water as received in the household was described by TW and TWA drinkers in Figure 4.39. Seventy-nine percent of TW drinkers described the taste of tap water as good (i.e., fresh, or no perceived taste) versus 56% of TWA drinkers. Twenty-two percent of TWA drinkers perceived tap water as having chlorinous flavors versus 4% of TW drinkers. A low percentage of both groups described tap water taste as metallic (2 to 4%). Two percent of TWA drinkers described tap water taste as being state/flat and musty/earthy, while these categories were not cited by TW drinkers.

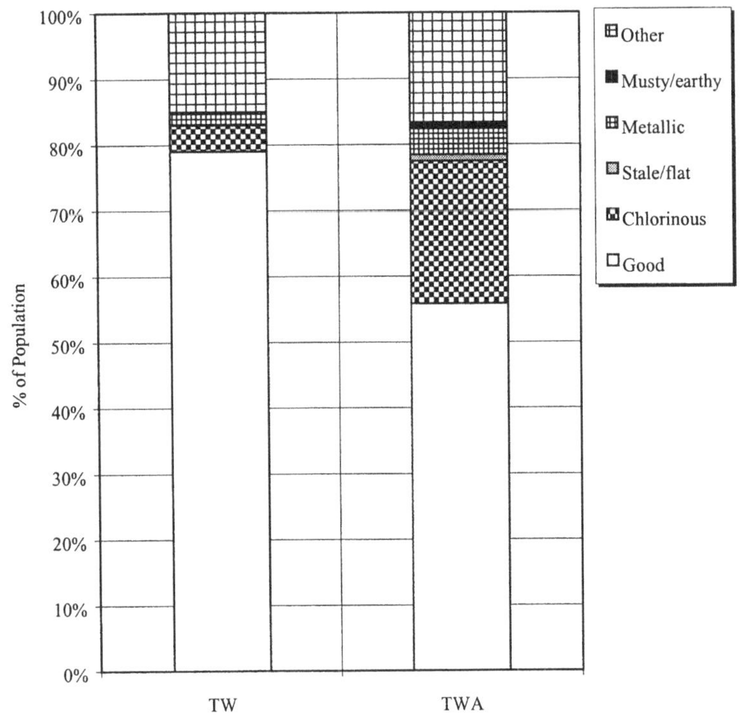

Figure 4.39 Description of tap water taste as perceived by TW and TWA drinkers - East 2 market (Q045)

Consumer Perception of Overall Utility Quality

The East 2 market consumers described their overall satisfaction with their local utilities. Eighty-five percent of TW drinkers were satisfied to very satisfied, as compared to 57% of TWA drinkers. When asked about their satisfaction with resolution of complaints, TW drinkers were split equally between being very satisfied, satisfied, dissatisfied, and very dissatisfied. The majority (75%) of TWA drinkers opinions ranged from neutral to very satisfied (out of 12 total complaints reported), while the remainder were very dissatisfied with handling of complaints by their local provider.

Market Southeast 1

Physical Data

This Southeast municipality relies on an internationally acclaimed water reuse program to sustain water demand for a county of nearly 1 million residents. Working closely with regional wastewater treatment facilities, Southeast 1 draws from a regional aquifer that is recharged by reused wastewater applied at local orchards and golf courses. Thirty million gallons of treated wastewater recharge the aquifer each day.

While deep wells are usually considered protected from influence of surface activity, turbidity levels in these deep wells can rise considerably from its typical baseline of less than 1 NTU (Figure 4.40). While organic carbon levels are moderate, averaging 2.5 mg/L, the sole treatment process of chlorination at wellheads tends to result in above-average total trihalomethanes (TTHMs). Twenty percent of the wellheads had TTHM levels exceeding 70 µg/L in 1999. In part, this may be due to the high chlorine dose (11 mg/L) required to maintain a residual in the distribution system.

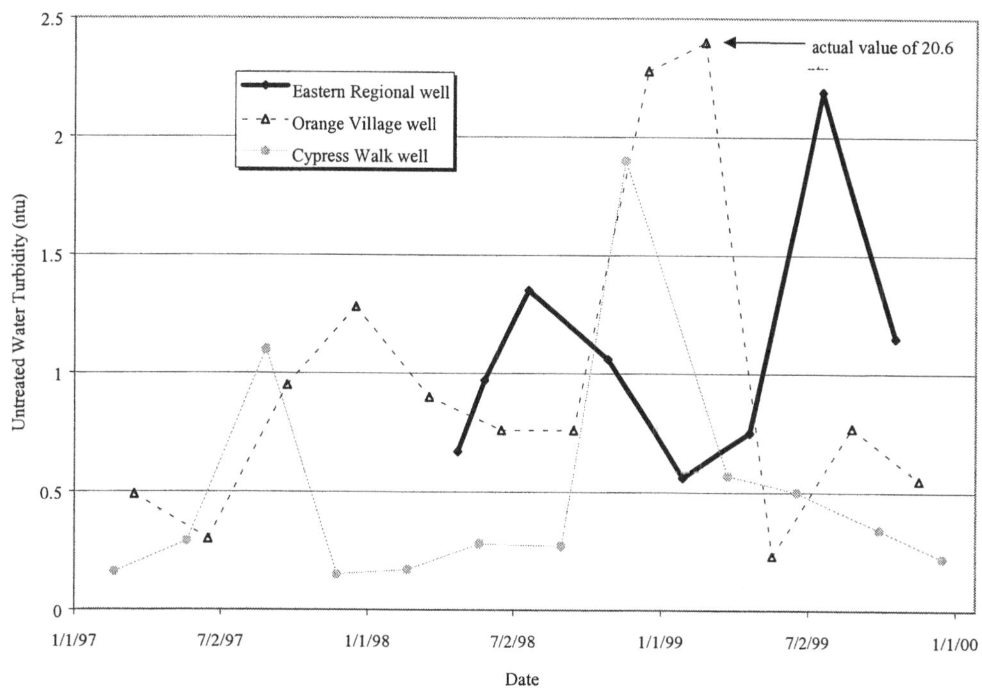

Figure 4.40 Fluctuation in turbidity for 3 wells - Southeast 1 market

Consumer Perception of Water Quality

Perceptions of Southeast 1 TW drinkers and TWA drinkers of their tap water quality features are illustrated in Figures 4.41 and 4.42. Regarding satisfaction with the overall quality of tap water, only 64% of TW drinkers were satisfied to very satisfied, as compared to 56% of TWA drinkers. Similar trends in perceptions were observed in both groups of consumers (i.e., TW drinkers and TWA drinkers). High proportions of satisfied consumers were observed in the smell and safety features of tap water (82 to 88% for TW drinkers and 71 to 72% for TWA drinkers). Fewer satisfied consumers were observed in the taste feature of tap water for both groups (63% for TW group and 45% for TWA group). Overall, there were more very dissastified consumers as compared to the other markets (1–7% for TW drinkers as compared to 5–18% for TWA drinkers).

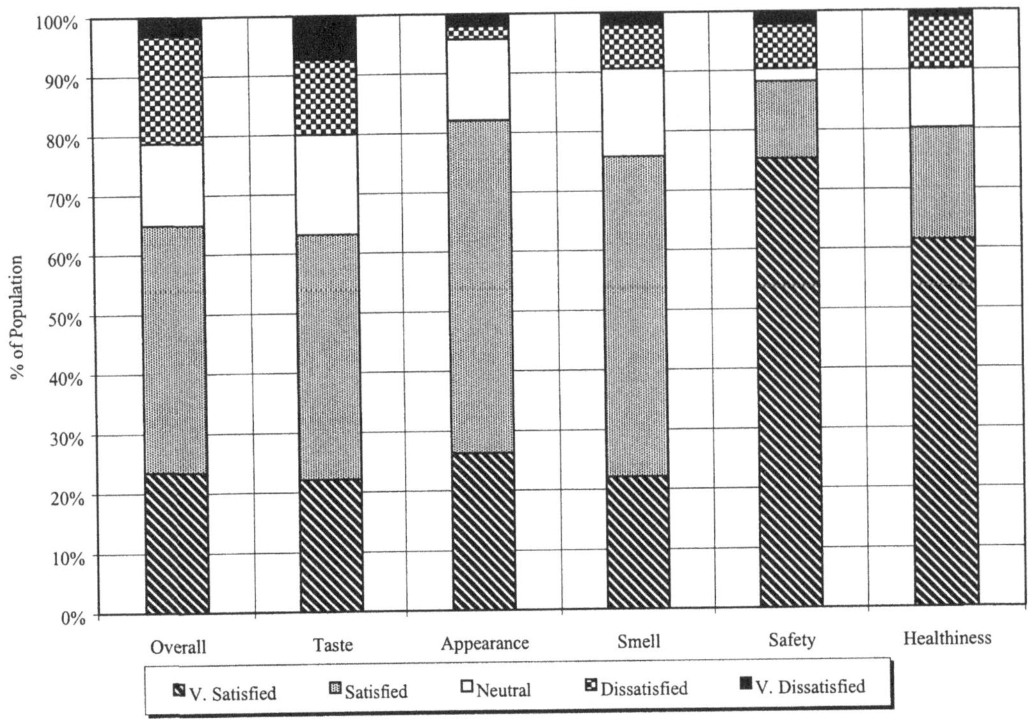

Figure 4.41 TW drinkers' satisfaction levels for various tap water quality issues - Southeast 1 market (Q005, Q010, Q015, Q020, Q025, Q026)

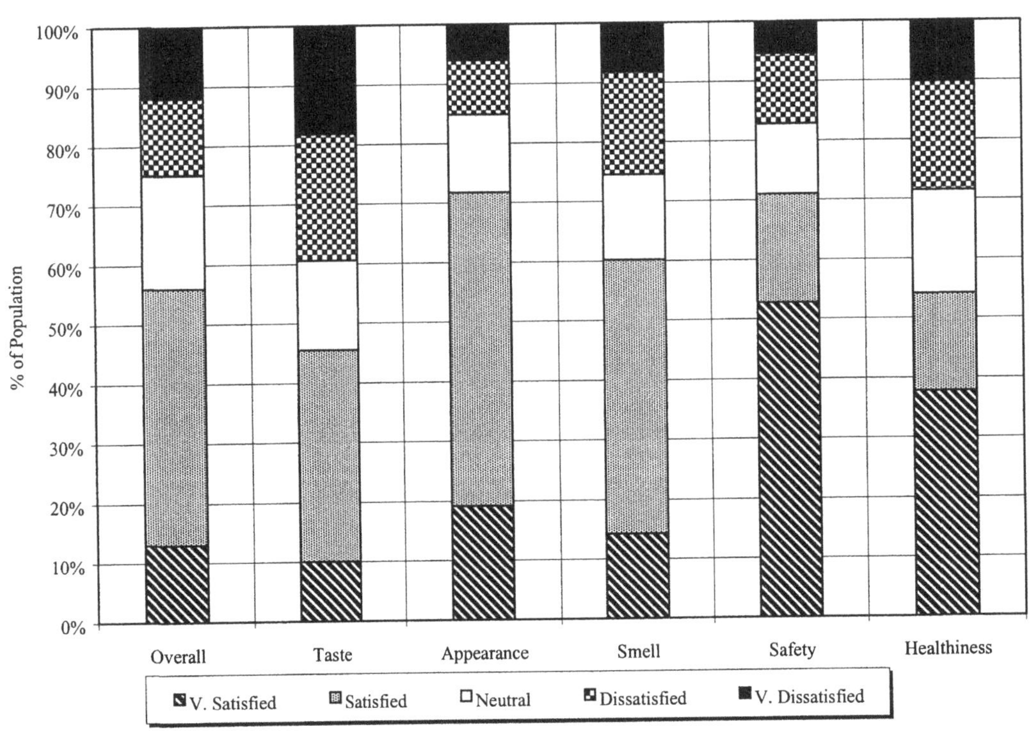

Figure 4.42 TWA drinkers' satisfaction levels for various tap water quality issues - Southeast 1 market (Q005, Q010, Q015, Q020, Q025, Q026)

BW and FW drinkers were asked to cite the most important reason they used tap water alternatives. The results are illustrated in Figure 4.43. Out of the tap water quality features of taste, appearance, smell, safety, and overall quality, tap water safety ranked as the number one reason for FW drinkers (55%) followed by taste (27%), and overall quality (18%). Out of the overall tap water quality features of taste, appearance, smell, healthiness, safety, and overall quality, tap water healthiness (35%) ranked as the number one reason for BW drinkers followed by taste (31%), smell (23%), safety (8%), and appearance (4%). These observations matched the TWA group's level of satisfaction in tap water quality features.

The taste of tap water as received in the household was described by TW and TWA drinkers in Figure 4.44. Sixty-eight percent of TW drinkers described the taste of tap water as good (i.e., fresh or no perceived taste) versus 50% of TWA drinkers. Twenty-four percent of TWA drinkers perceived tap water as having chlorinous flavors versus 7% of TW drinkers.

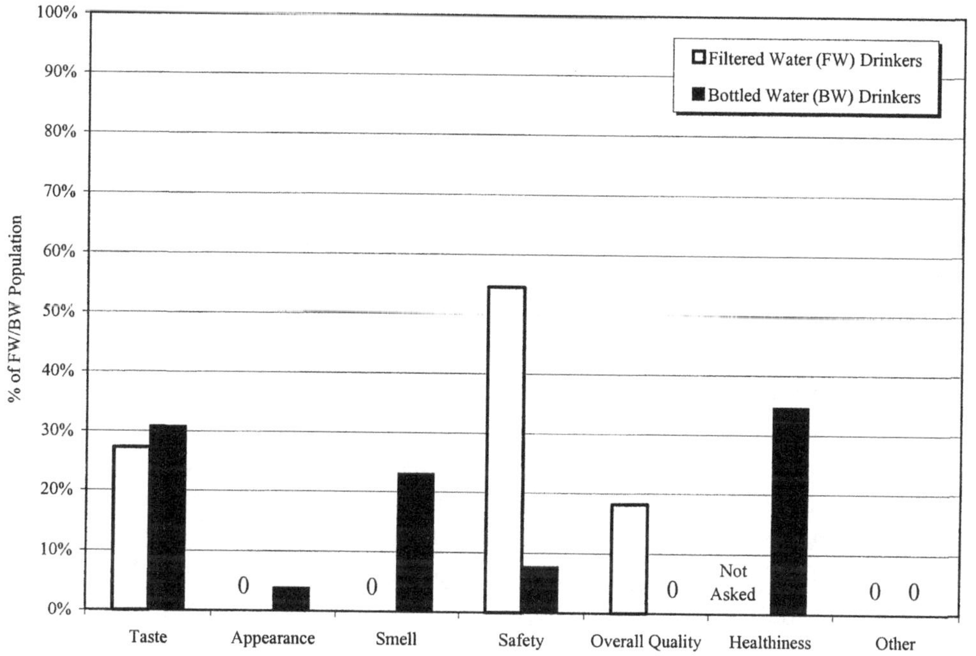

Figure 4.43 Most important reasons for drinking tap water alternatives as cited by BW and FW drinkers -Southeast 1 market (Q091, Q102)

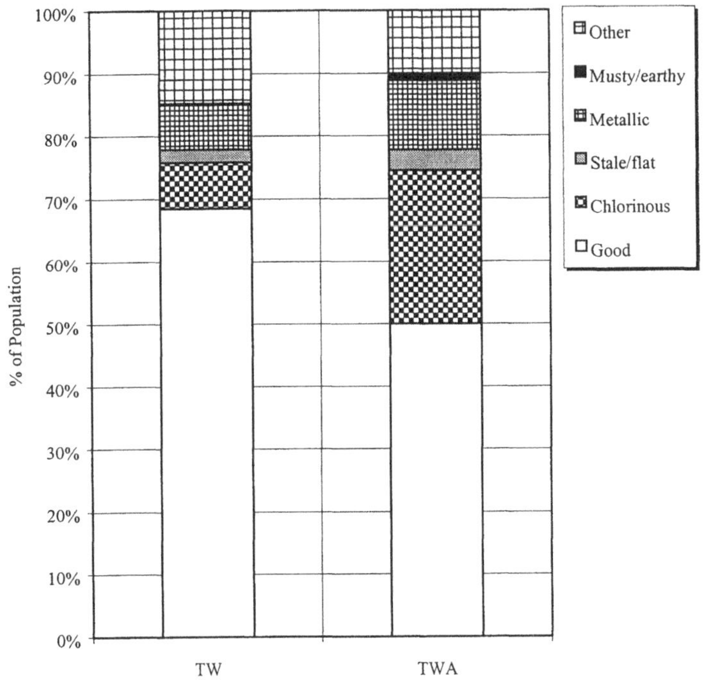

Figure 4.44 Description of tap water taste as perceived by TW and TWA drinkers - Southeast 1 market (Q045)

Consumer Perception of Overall Utility Quality

The Southeast 1 market consumers from both groups described their overall satisfaction with their local utility. Seventy-five percent of TW drinkers were satisfied to very satisfied as compared to 60% of TWA drinkers. When asked about their satisfaction with resolution of complaints, only 40% of TW group (16 complaints reported), and 38% of TWA group (14 complaints reported) were satisfied to very satisfied with the complaint resolution of their local provider.

Market Southeast 2

Physical Data

Southeast 2 is located on the Gulf of Mexico and provides service to nearly 300,000 citizens. Recurring water shortage problems have prompted the utility to evaluate brackish or seawater alternatives, but a less costly alternative, a 100-mile pipeline, proved to be the final solution. Consumer complaints increased once the pipeline was online and it was thought that the problem (color) was due to source water characteristics. Further examination revealed that the color/turbidity issue was linked with a change in the treatment process, which occurred concurrently with the pipeline installation. With time, a protective scale layer on the new pipe and process changes helped alleviate the color/turbidity problem.

Water for Southeast 2 can be classified as a hard, mineral water. Treated water has a salt content of approximately 480 mg/L, hardness of 180 mg/L as $CaCO_3$ and alkalinity of 140 mg/L as $CaCO_3$. Noticeable water quality fluctuations can occur for this surface water facility, as observed during heavy rain in August 1999 (Figure 4.45). In addition to a TOC increase, turbidity rose to 200 NTU during this time. While the use of chloramines as a secondary disinfectant has alleviated chlorine taste and odor complaints, the high finished water temperatures of 27°C are a common complaint.

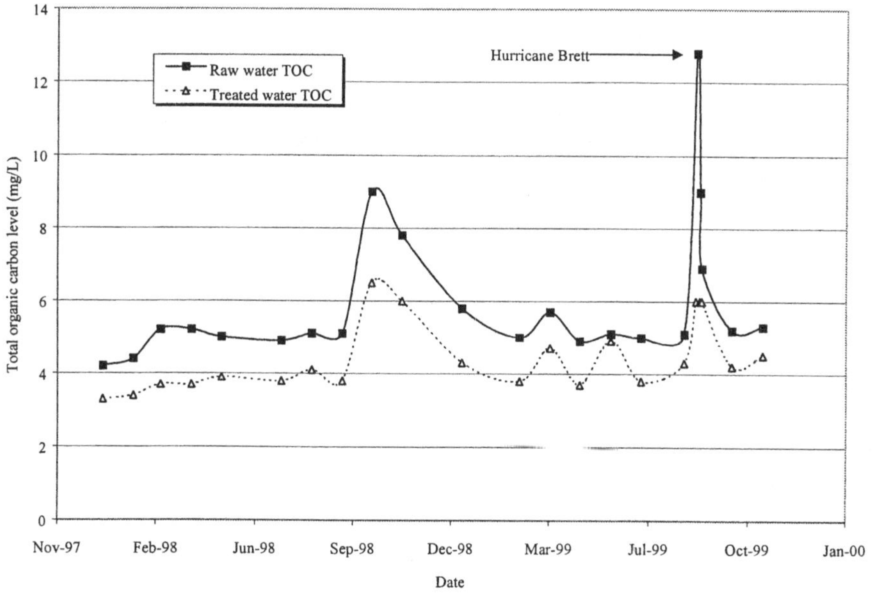

Figure 4.45 TOC variability and removal efficiency - Southeast 2 market

Consumer Perception of Water Quality

Perceptions of Southeast 2 TW drinkers and TWA drinkers of their tap water quality features are illustrated in Figures 4.46 and 4.47. Regarding satisfaction with the overall quality of tap water, 95% of TW drinkers were satisfied to very satisfied, as compared to 75% of TWA drinkers. Also, regarding satisfaction with appearance, smell, safety, and healthiness features, 89 to 95% of TW drinkers were satisfied to very satisfied, as compared to 85% for FW drinkers. Overall, there were very few consumers who were very dissatisfied with any of the six tap water quality features (0% for TW drinkers as compared to 1–5% for TWA drinkers).

BW and FW drinkers were asked to cite the most important reason they used tap water alternatives. The results are illustrated in Figure 4.48. Out of the tap water quality features of taste, appearance, smell, safety, and overall quality, tap water safety ranked as the number one reason for FW drinkers (57%), followed by taste (38%), and overall quality (7%). Out of the overall tap water quality features of taste, appearance, smell, healthiness, safety, and overall quality, tap water healthiness (41%) ranked as the number one reason for BW drinkers followed by safety (25%), taste (19%), and overall quality (8%).

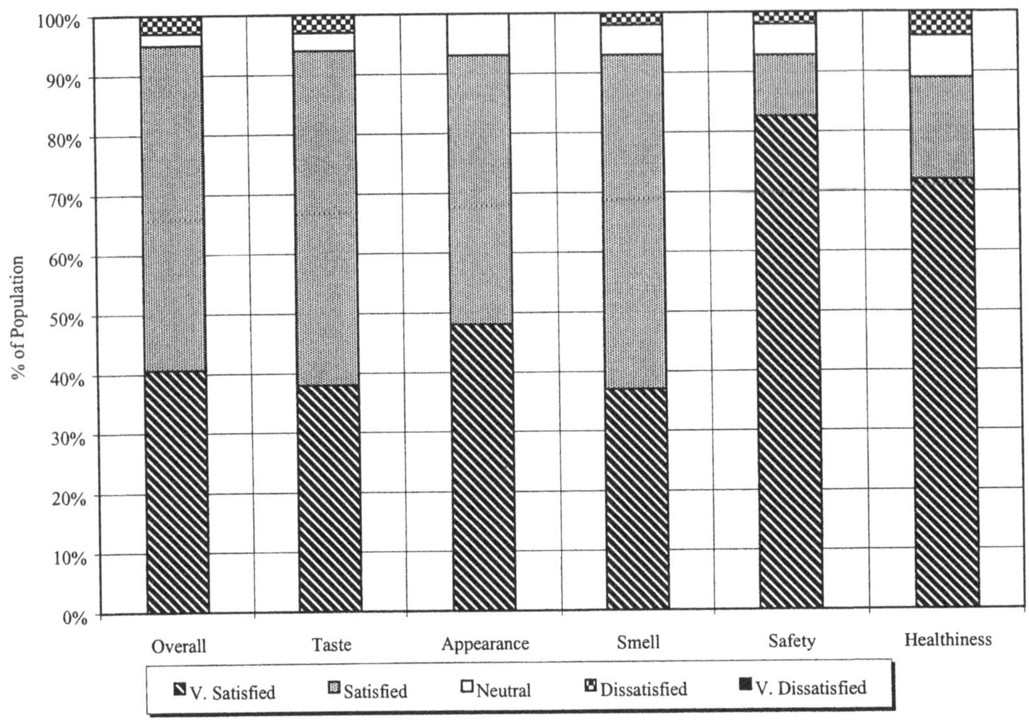

Figure 4.46 TW drinkers' satisfaction levels for various tap water quality issues - Southeast 2 market (Q005, Q010, Q015, Q020, Q025, Q026)

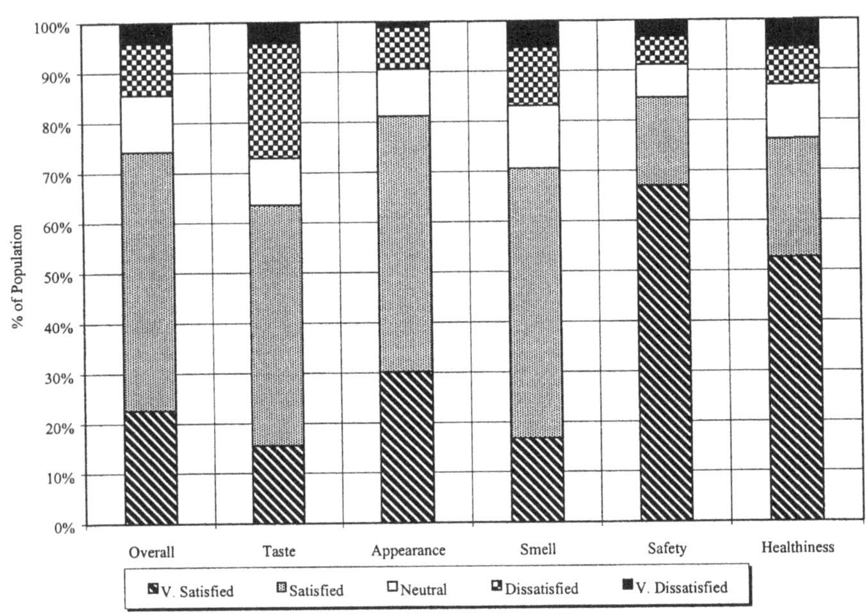

Figure 4.47 TWA drinkers' satisfaction levels for various tap water quality issues - Southeast 2 market (Q005, Q010, Q015, Q020, Q025, Q026)

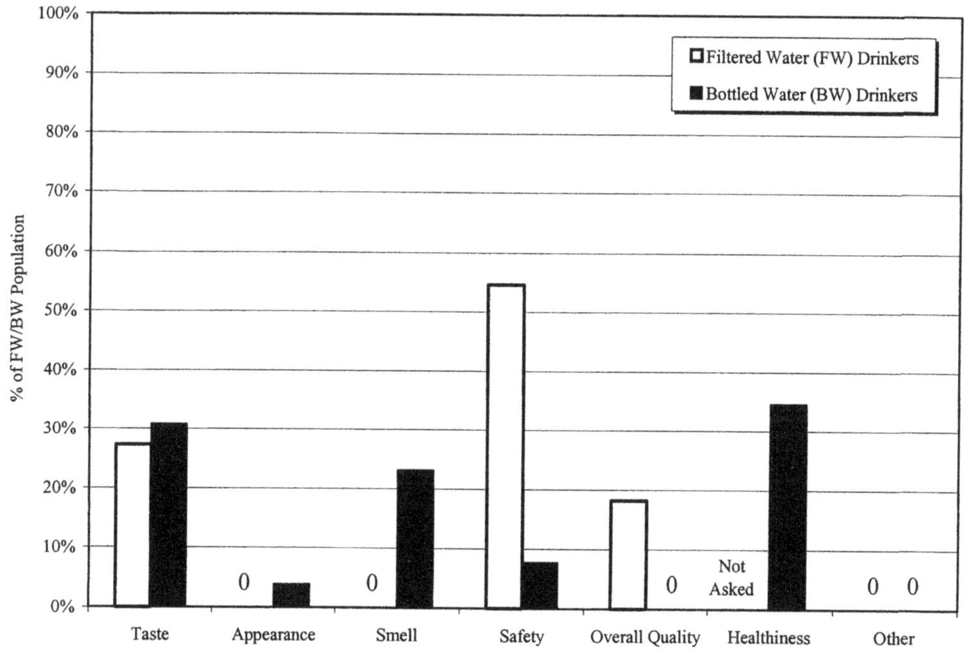

Figure 4.48 Most important reasons for drinking tap water alternatives as cited by BW and FW drinkers - Southeast 2 market (Q091, Q102)

The taste of tap water as received in the household was described by TW and TWA drinkers in Figure 4.49. Seventy-two percent of TW drinkers described the taste of tap water as good versus 44% of TWA drinkers. Thirty-three percent of TWA drinkers perceived tap water as having chlorinous flavors versus 6% of TW drinkers. A low percentage of both groups described tap water taste as metallic (0 to 2%). One percent of TWA drinkers described tap water taste as being musty/earthy, while TW drinkers did not report tastes in this category.

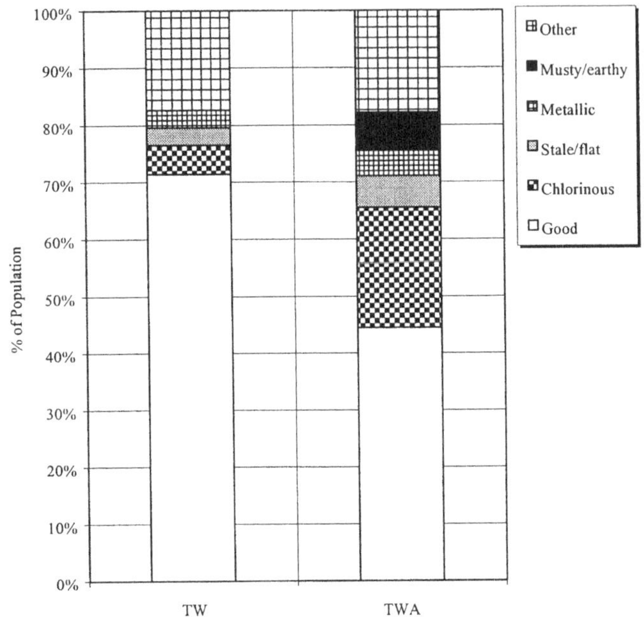

Figure 4.49 Description of tap water taste as perceived by TW and TWA drinkers - Southeast 2 market (Q045)

Consumer Perception of Overall Utility Quality

Southeast 2 consumers from both groups described their overall satisfaction with their local utility. Seventy-nine percent of TW drinkers were satisfied to very satisfied as compared to 59% of TWA drinkers. When asked about their satisfaction with resolution of complaints, 37% of TW drinkers (out of 11 complaints reported) were very satisfied to satisfied, while 50% of TWA drinkers (out of 8 complaints reported) were satisfied with resolution of complaints by their local provider.

Market South

Physical Data

The South market supplies water to its consumers from three reverse osmosis (RO) water treatment plants and supplements it's RO capacity with water purchased from a nearby city. Recent drops in the aquifer level initiated an Aquifer Storage and Recovery (ASR) system and helped to guarantee a future supply of water. An advanced treatment technology, RO, is required due to high level of TDS and fluoride in the local aquifer.

Finished water quality consists of TDS and hardness levels of 100 or 268 mg/L (depending upon the source) and 33 - 55 mg/L as $CaCO_3$ (depending upon the source), respectively. The main aesthetic challenge faced by this utility is warm water temperatures, which can reach a hot 32°C during the summer. In spite of this, consumer surveys handed out by the utility indicated that the respondents were satisfied with overall quality (69%), safety/healthiness (90%), taste (83%), and smell (89%).

Consumer Perception of Water Quality

Perceptions of the South market TW drinkers and TWA drinkers of their tap water quality are illustrated in Figures 4.50 and 4.51. Regarding satisfaction with the overall quality of tap water, 93% of TW drinkers were satisfied to very satisfied, as compared to 66% of TWA drinkers. Similar trends in perceptions were observed in both groups of consumers (i.e., TW drinkers and TWA drinkers High proportions of satisfied consumers were observed in appearance, smell, and safety features (93 to 96% for TW drinkers and 83–89% for TWA drinkers). Fewer satisfied consumers were observed in the taste feature of tap water for the TWA group (59% as compared to 91% for the TW group). Overall, there were very few consumers who were very dissatisfied with any of the six tap water quality features (0–3% for TW drinkers as compared to 1–7% for TWA drinkers).

BW and FW drinkers were asked to cite the most important reason they used tap water alternatives. The results are illustrated in Figure 4.52. Out of the tap water quality features of taste, appearance, smell, safety, and overall quality, tap water safety ranked as the number one reason for FW drinkers (57%), followed by taste (27%), and overall quality (13%). Out of the overall tap water quality features of taste, appearance, smell, healthiness, safety, and overall quality, tap water safety (32%) ranked as the number one reason for BW drinkers followed by taste (19%), overall quality (16%), and healthiness (13%). These perceptions exist despite the fact that 83% of TWA drinkers perceived tap water as being safe to drink.

The taste of tap water as received in the household was described by TW and TWA drinkers in Figure 4.53. Sixty-seven percent of TW drinkers described the taste of tap water as good versus 55% of TWA drinkers. Seven percent of TWA drinkers perceived tap water as having chlorinous flavors versus 1% of TW drinkers.

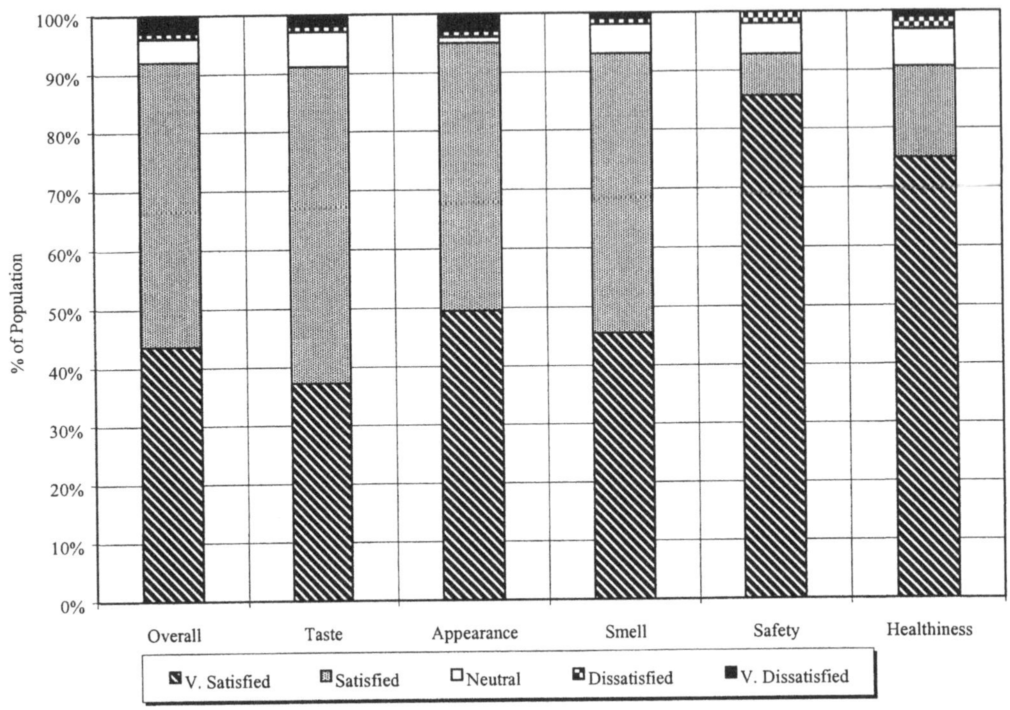

Figure 4.50 TW drinkers' satisfaction levels for various tap water quality issues - South market (Q005, Q010, Q015, Q020, Q025, Q026)

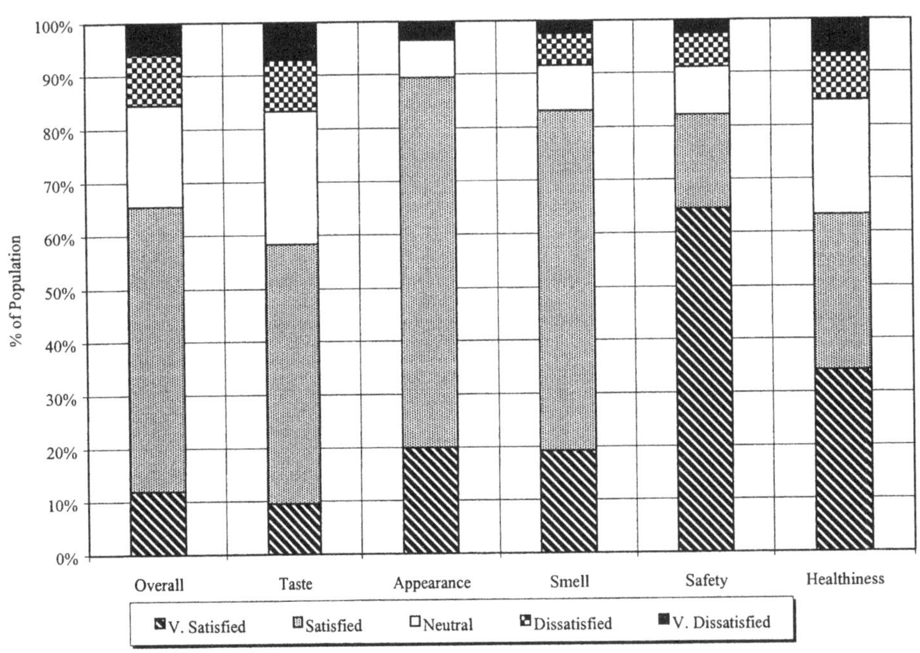

Figure 4.51 TWA drinkers' satisfaction levels for various tap water quality issues - South market (Q005, Q010, Q015, Q020, Q025, Q026)

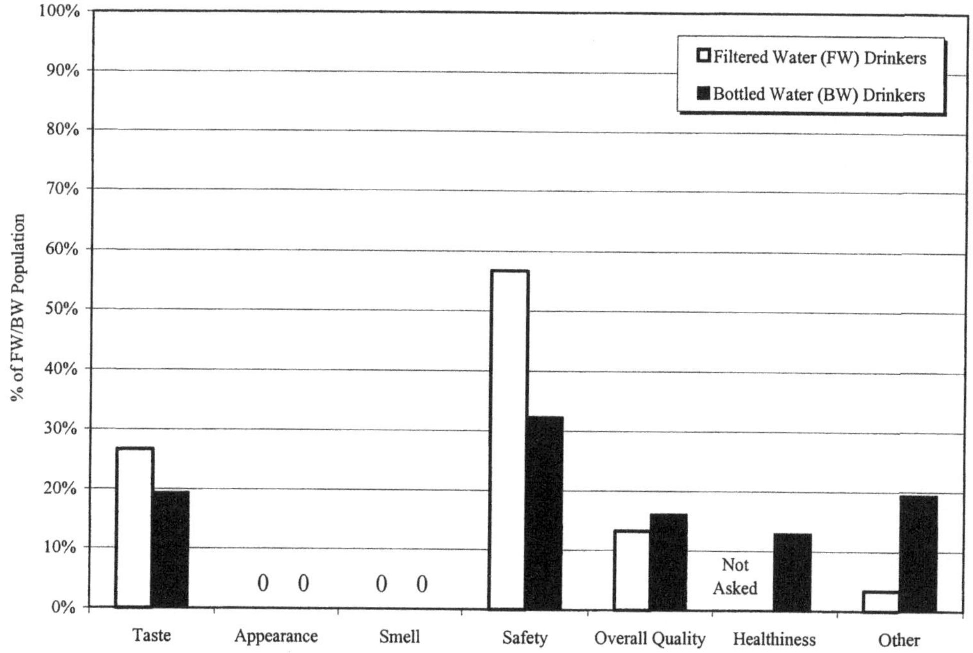

Figure 4.52 Most important reasons for drinking tap water alternatives as cited by BW and FW drinkers - South market (Q091, Q102)

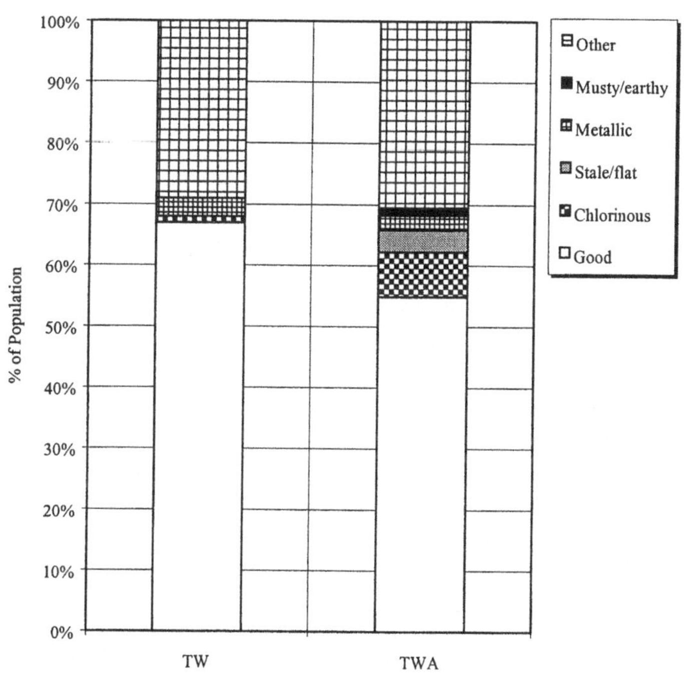

Figure 4.53 Description of tap water taste as perceived by TW and TWA drinkers - South market (Q045)

Consumer Perception of Overall Utility Quality

The South market consumers from both groups described their overall satisfaction with their local utility. Ninety-eight percent of TW drinkers were satisfied to very satisfied as compared to 76% of TWA drinkers. When asked about their satisfaction with resolution of complaints, 57% of TW drinkers (14 complaints total) and 41% of TWA drinkers (17 complaints total) were satisfied to very satisfied with the service of their local provider.

CHAPTER 5
CORRELATIONS AND DIFFERENCES BETWEEN MARKETS

In this chapter, market comparisons of TWA drinkers based on demographics, satisfaction levels, and utility characteristics are examined. All differences between consumer groups were tested using one-way ANOVA tests or Chi-squared analyses at a 95% level of confidence. Discussions are grouped into four sections:

- Distribution of TWA drinkers by market
- Demographic induced differences affecting tap water satisfaction and use of tap water alternatives
- Utility induced differences affecting tap water satisfaction and use of tap water alternatives
- Correlation between consumer satisfaction and water quality/utility characteristics

For demographic differences markets were combined into 7 geographically similar regions (Fig. 5.1). To determine differences based on utility characteristics each market was kept separate (Fig. 5.2).

Tap water alternative use in markets across the U.S. ranged from a low of 20% (Midwest 3) to a high of 87% (West Coast). Overall, 48% of total respondents consume primarily tap water alternative products, though significant geographic variability results in a large deviation from this average (Figure 5.3). In markets with a significant number of either TW or TWA drinkers, quota cells of 100 respondents were not always filled (see Table 2.3). In the Southeast 1 market, 371 TWA drinkers were surveyed before obtaining 95 TW drinkers. At the opposite extreme in the Midwest 3 market, the market with the lowest consumption of TWAs (20%), 381 TW drinkers werc surveyed before obtaining 96 TWA drinkers. Sufficient respondents were contacted in each market (at least 350 people per market) to determine current consumption rates for TWA drinkers in each market.

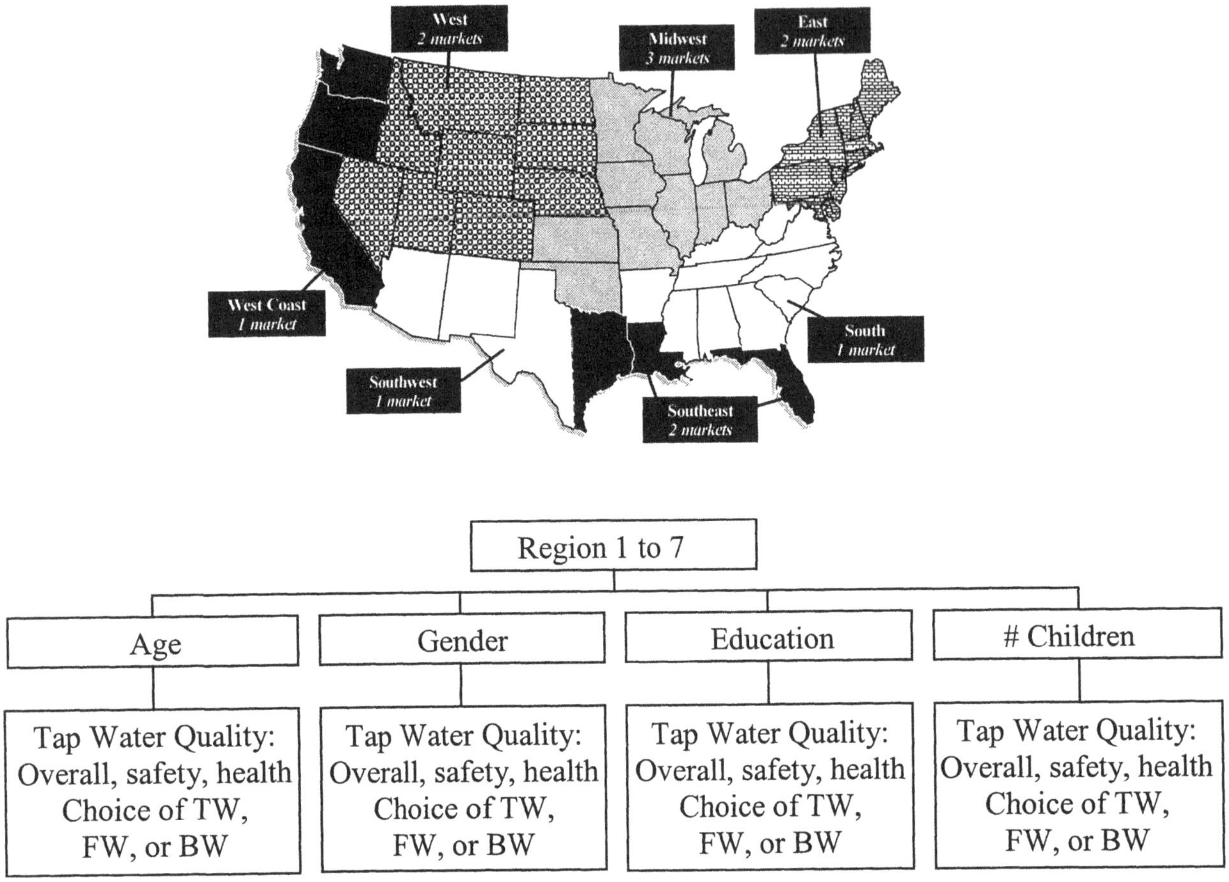

Figure 5.1 Regional analysis of demographics' effect on satisfaction of tap water quality and choice of drinking water source

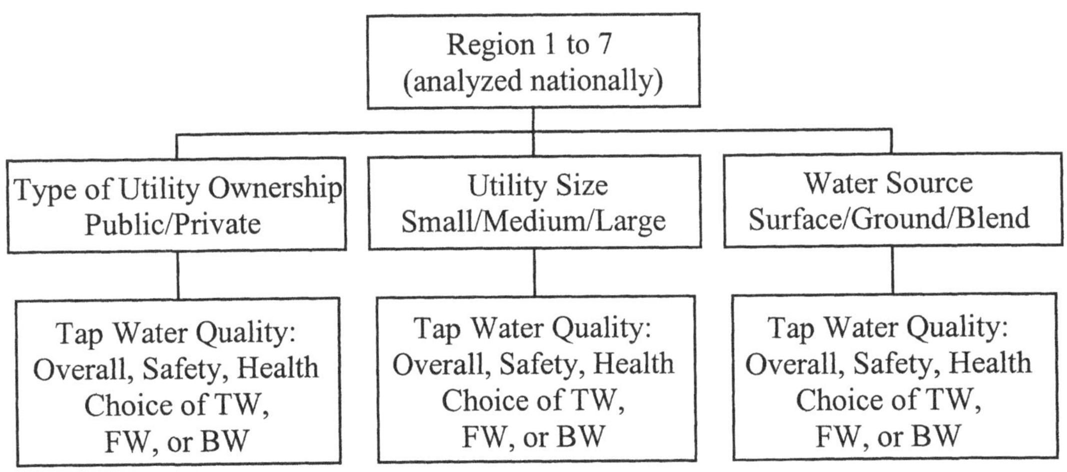

Figure 5.2 Regional analysis of effect of utility details on perception of tap water quality and choice of drinking water source

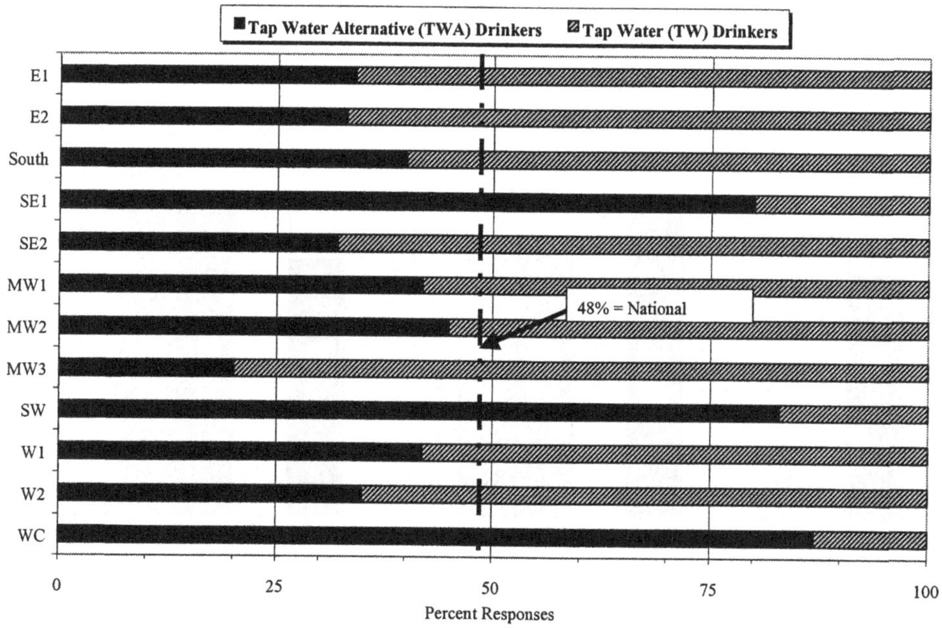

Figure 5.3 Projected use of tap water and tap water alternatives for the 12 market areas (QuotaQ)

DEMOGRAPHIC DIFFERENCES BASED ON WATER QUALITY SATISFACTION

Responses to the following questions were analyzed by age, gender, education level, number of children, type of utility ownership, utility size, and water source:

- What is your satisfaction with the overall quality of the tap water you receive in your household (Q005)?
- How safe is the tap water you receive in your household (Q025)?
- How healthy is the tap water you receive in your household (Q026)?
- Which water quality feature is most important to you (taste, appearance, smell, safety, healthfulness) (Q030)?
- Do you use a water filtration system for the tap water you receive in your household (Q065)?
- Do you drink bottled water while at home (Q095)?

Demographic comparisons based on tap water alternative use are shown in Figures 5.4 and 5.5. Results from all 12 markets were combined. For each category in a given demographic, the percent TWA drinkers is displayed; the remainder (i.e., 100% - listed value) is made up of TW drinkers.

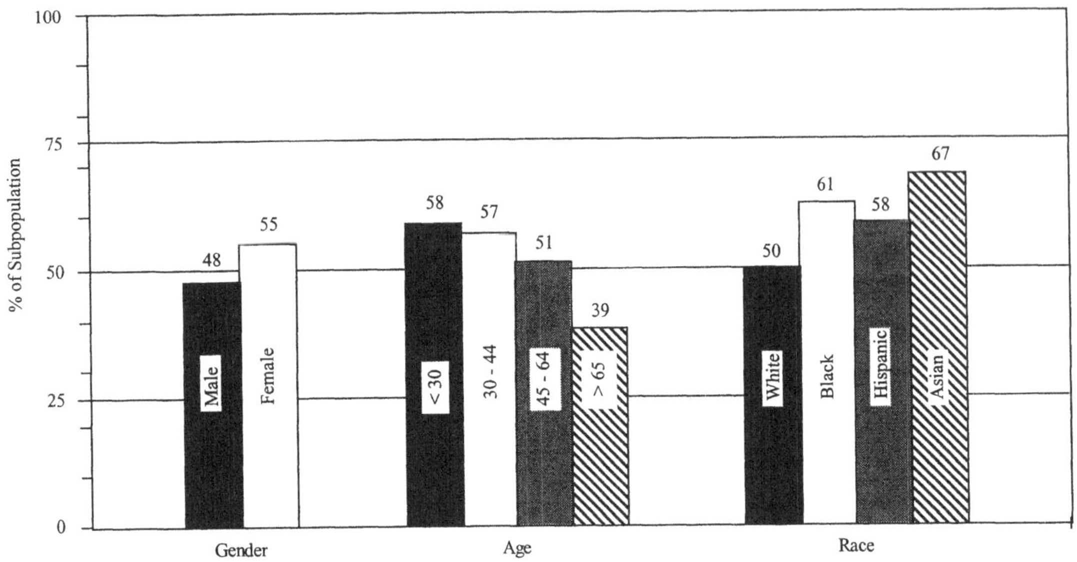

Figure 5.4 TWA drinkers' demographic composition (percentages represent TWA drinkers of a particular subpopulation)

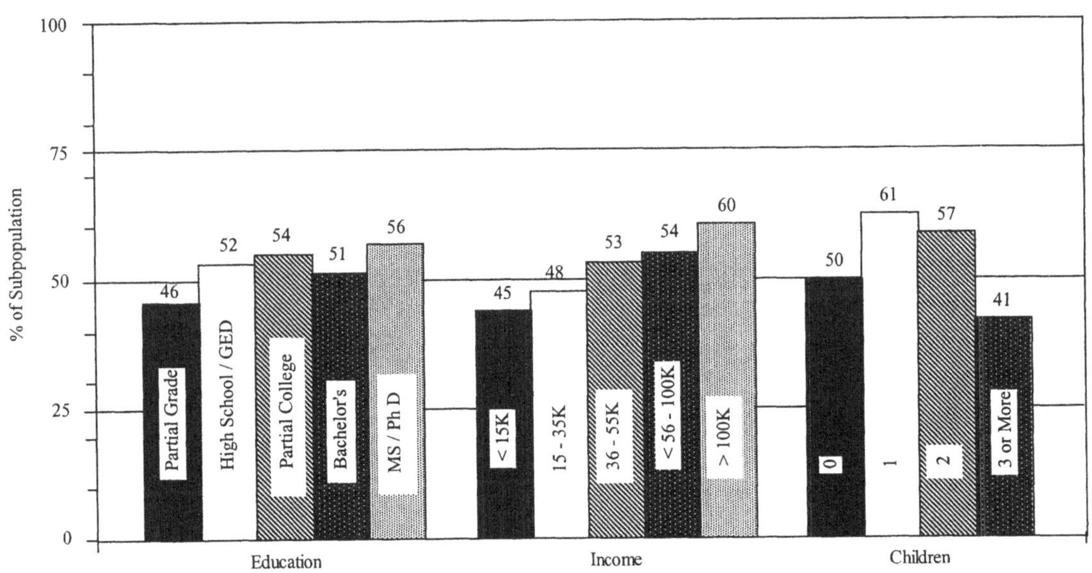

Figure 5.5 TWA drinkers' demographic composition (percentages represent TWA drinkers of a particular subpopulation)

Tests for differences (one-way ANOVA test at the 95% confidence level) in perception among subpopulations of TW features in each region are reported in tabular format. If a particular cell has no entry, there was no difference in perception based upon the specific demographic. For instances where a statistically significant difference exists, the exact relationship is written out in shorthand. See table footnotes for explanations of abbreviations used in each table.

Age was an important factor in the satisfaction of the overall quality of the tap water in the Southwest, West, Midwest, East Coast, and Southeast (Table 5.1). Across these markets consumers younger than 30 were less satisfied in the overall quality of tap water than were consumers who were 65-years-old or older. Consumers between and including the ages of 30 and 64 were more satisfied than consumers younger than 30 and were less satisfied than those 65 and older. In the Midwest, households with children were less satisfied with overall tap water quality than households without children. In general, education and gender did not affect the average level of satisfaction with tap water quality.

Table 5.1
Regional analysis of effect of demographics on perception of overall tap water quality

Geographical region	Market	Age	Gender	Education	Children
West Coast	West Coast				
Southwest	Southwest	Under 30<30 to 44 &45 to 64&65 or older			
West	West 1	Under 30<65 or older			
	West 2				
Midwest	Midwest 1	Under 30 & 30 to 44<65 or older			
	Midwest 2				HH w/child<HII w/o child
	Midwest 3	30 to 44 & under 30<65 or older			HH w/child<HH w/o child
East Coast	East 1	Under 30 and 45-64<30-44 and 65 or older			
	East 2	Under 30<65 or older			
Southeast	Southeast 1	Under 30<45-64			
	Southeast 2		M>F	Associates degree< Graduate School	
South	South				

HH Household M Male F Female

Age also had the most impact on the perception of tap water safety (Table 5.2). Relative to consumers 65 and older, consumers 44 and younger perceived their tap water as being less safe to drink. In Southeast 2 market, households with children perceived tap water as less safe than did households without children, and people with their associate's degree perceived tap water as being less safe than did people with a graduate degree.

Table 5.2
Regional analysis of effect of demographics on perception of safety of tap water

Geographical region	Market	Age	Gender	Education	Children
West Coast	West Coast				
Southwest	Southwest				
West	West 1	Under 30<45 to 64 and 65 or older			
	West 2				
Midwest	Midwest 1	30 to 44<65 or older			
	Midwest 2				
	Midwest 3	Under 30<30 to 44 &45 to 64 and 65 or older			
East Coast	East 1	Under 30 and 45-64<65 or older			
	East 2				
Southeast	Southeast 1	30 to 44<65 or older			
	Southeast 2				HH with child<HH w/o child
South	South				

Differences in the perception of tap water healthiness are illustrated in Table 5.3. Once again, age of the respondents was a significant factor. Consumers aged 45 to 69 perceived their tap water as being less healthy relative to consumers 65 and older. In four of the seven regions, the presence of children in the household was another influential factor. In these regions, households with children perceived their tap water as being less healthy than did households without children.

Table 5.3

Regional analysis of effect of demographics on the perception of tap water healthiness

Geographical region	Market	Age	Gender	Education	Children
West Coast	West Coast	30 to 44& under 30&45-64<65 or older			
Southwest	Southwest	under 30<65 or older			
West	West 1				
	West 2				
Midwest	Midwest 1		F>M		
	Midwest 2	under 30 & 30 to 44 < 65 or older			HH with child<HH w/o child
	Midwest 3	under 30 <30 to 44 &45 to 64 and 65 or older			HH with child<HH w/o child
East Coast	East 1	under 30<65 and older			
	East 2	Under 30 & 30-34 < 65 or older			HH with child<HH w/o child
Southeast	Southeast 1	Under 30 & 30-34 < 65 or older			
	Southeast 2	30 to 45<65 or older			HH with child<HH w/o child
South	South				

HH Household M Male F Female

An analysis of the most important tap water quality for consumers revealed only a few statistically significant differences between demographic groups (Table 5.4). In the West 2 market, consumers under the age of 30 were more likely to cite smell and less likely to cite healthiness as being the most important feature. In the East 2 market, male respondents were more likely to indicate taste, whereas female respondents were more likely to indicate safety. In the South, smell and healthiness were the more likely choices for students with 1-3 years of College and with an Associate's Degree. Safety was the most likely feature chosen by households with children in the South, whereas people in households without children were more likely to cite healthiness.

Table 5.4
Regional analysis of effect of demographics on the perception of most important tap water quality feature

Geographical region	Market	Age	Gender	Education	Children
West Coast	West Coast				
Southwest	Southwest				
West	West 1				
	West 2	Under 30 ML smell; under 30 LL indicate health			
Midwest	Midwest 1				
	Midwest 2				
	Midwest 3				
East Coast	East 1				
	East 2		M ML indicate taste, F ML indicate safety		
Southeast	Southeast 1				
	Southeast 2				
South	South			college 1-3 yrs ML indicate health; 2-years Associates Degree ML indicate smell	HH with children ML indicate safety; HH w/o children ML indicate healthiness

HH Household M Male F Female LL Least likely ML Most likely

Age was also a factor in consumers' choice to use a water filtration system in their household (Table 5.5). In the Southwest, the 45 to 64 year old consumers were more likely to use a filtration system, while 30 to 44 year old consumers in the South were more likely to use a filtration system. Southern consumers between the ages of 45 and 64 were more likely not to use a filtration system. In the Southeast, it was generally observed that consumers with advanced education were more likely to utilize a water filtration system in their home than those without 4-year college or graduate degrees.

Table 5.5
Regional analysis of correlation between demographics and choice to use a Point of Use device to filter the tap water received in the household

Geographical region	Market	Age	Gender	Education	Children
West Coast	West Coast				
Southwest	Southwest	45 to 64 ML use FW			
West	West 1		F ML use FW, M LL use FW		
	West 2				
Midwest	Midwest 1				
	Midwest 2		M ML own		
	Midwest 3				
East Coast	East 1				
	East 2				
Southeast	Southeast 1			College 4 yrs or more ML use FW; Grade 12 or GED LL use FW	
	Southeast 2				
South	South	30-44 ML use FW; 45-64 ML not use FW			

HH Household M Male F Female LL Least likely ML Most likely

In six of the seven geographic regions, age was an important factor in consumers' decisions to drink bottled water while at home (Table 5.6). Consumers who were 44 years-old or younger were more likely to drink bottled water at home, while consumers 65 and older were less likely to drink bottled water at home. Another factor affecting the choice to drink bottled water was the presence of children in the household. Households with children in the West, Midwest, and East Coast regions were more likely to drink bottled water while at home.

Table 5.6
Regional analysis of correlation between demographics and choice to drink bottled water while at home

Geographical region	Market	Age	Gender	Education	Children
West Coast	West Coast				
Southwest	Southwest	under 30 ML drink BW, 65 or older LL drink BW		College 1-3 years ML drink BW; College >4 years LL drink BW	
West	West 1	30 to 44 ML drink BW, 65 or older LL drink BW			HH with child ML drink BW
	West 2				HH with child ML drink BW
Midwest	Midwest 1	30 to 44 ML drink BW, 65 or older LL drink BW			
	Midwest 2				HH with child ML drink BW
	Midwest 3	under 30 & 30 to 44 ML drink BW; 65 or older LL drink BW			HH with child ML drink BW
East Coast	East 1	under 30 ML drink BW; 65 or older LL drink BW			
	East 2	under 30 ML drink BW, 65 or older LL drink BW			HH with child ML drink BW
Southeast	Southeast 1	65 or older LL drink BW			
	Southeast 2	30 to 44 ML drink BW, 65 or older LL drink BW			
South	South	65 or older LL drink BW			

HH Household; M Male; F Female; LL Least likely
ML Most likely; FW Filtered water; BW Bottled water

TAP WATER USAGE AND SATISFACTION BASED ON UTILITY CHARACTERISTICS

Table 5.7 describes the ownership, size, and raw water source of each utility surveyed. Each of these utility characteristics was analyzed with consumers' satisfaction with their tap water quality and tap water provider (Table 5.8). Consumers using public water providers were more confident in the safety of their tap water than were consumers using private water providers. Consistent with this result, those served by private water providers were more likely to use a filtration device or to drink bottled water while at home. Additionally, consumers served by public water providers were 1) more confident that their tap water meets state and federal regulations, 2) more satisfied with their local water provider, and 3) more likely to be aware of consumer education programs initiated by their local provider.

Table 5.7
Classification of each market in terms of water provider (WP) ownership type, size, and water source

Market area	WP type	WP size	WP source
West Coast	Private	Moderate	Ground water
Southwest	Public	Moderate	Surface water
West 1	Private	Moderate	Ground water
West 2	Public	Moderate	Surface water
Midwest 1	Public	Large	Surface water
Midwest 2	Public	Small	Surface water
Midwest 3	Public	Moderate	Surface water
East 1	Public	Large	Surface water
East 2	Private	Moderate	Surface water
Southeast 1	Public	Moderate	Ground water
Southeast 2	Public	Large	Surface water
South	Public	Small	Surface water

Small 0-5 MGD
Moderate 5-20 MGD
Large >50 MGD
* For blended systems, classification is based upon the dominant source (i.e., >50% of water used is surface water, then utility was classified as "surface water")

Table 5.8
National market analysis - correlations between utility characteristics and satisfaction with tap water quality and water provider

Question	Utility ownership	Utility size	Water source
Satisfaction with overall quality of tap water		Large and moderate<small	Ground<surface
Satisfaction with tap water safety	Public>private	Large and moderate<small	Ground<surface
Satisfaction with tap water healthiness		Large and moderate<small	Ground<surface
Most important water quality feature		Large WP size ML appearance, smell	Ground ML healthiness, Surface ML safety
Do you filter the tap water received in the household?	Those with private WP ML to use water filtration system	Large ML to not use water filter; Moderate ML to use filter	Ground ML to use water filtration system
Do you drink bottled water while at home?	Those with private WP ML to drink BW while at home		Ground ML to drink BW
How confident are you that tap water meets state and federal requirements?	consumers of public WP more confident than consumers of private providers	Consumers of large WP less confident than consumers of moderate WP. Consumers of moderate WP less confident than consumers of small WP	consumers of WP using GW less confident than consumers of WP using SW
Satisfaction with water provider	consumers of public WP more satisfied than consumers of private providers	Consumers of large WP less satisfied than consumers of moderate WP. Consumers of moderate WP less satisfied than consumers of small WP	consumers of WP using GW less satisfied than consumers of WP using SW
Satisfaction with water quality info provided by the water provider	consumers of public WP more satisfied than consumers of private providers	Consumers of large and moderate WP less satisfied than consumers of small WP	consumers of WP using GW less satisfied than consumers of WP using SW
Awareness of consumer education programs initiated by water provider	Those with public WP ML to be aware of consumer Ed programs	Consumers of small/large WP more likely to be aware of program. Consumers of moderate WP more likely to be unaware	consumers of WP using SW more likely to be aware of programs

WP Water provider   TW Tap water   BW Bottled water   SW Surface water   GW Groundwater

Consumers served by moderate to large water providers were less satisfied with overall tap water quality, safety, and healthiness relative to consumers of small water providers. These results were consistent with the observation that consumers of small water providers were 1) more likely to drink tap water at home, 2) more confident that their tap water meets state and federal requirements, and 3) more satisfied with their local provider. Consumers served by small and large water providers were more likely to be aware of consumer education programs while consumers served by moderate size water providers were more likely to be unaware of educational programs.

Consumers from water providers using groundwater were less satisfied with overall tap water quality, safety, and healthiness, relative to consumers from water providers using surface water. Accordingly, consumers from surface water providers were more likely to drink household tap water, whereas consumers from groundwater providers were more likely to use filtration devices or to drink bottled water while at home. Groundwater consumers were also 1) less confident that their tap water meets state and federal requirements, 2) less satisfied with their local provider, and 3) less likely to be aware of consumer education programs.

CORRELATIONS BETWEEN WATER QUALITY SATISFACTION AND WATER QUALITY/UTILITY CHARACTERISTICS

Differences in satisfaction levels and consumption patterns among demographic groups helped identify important subgroups in the population but did not provide an estimate of the degree of difference. Linear regression was used for scenarios where both dependent and independent variables could be expressed using ordinal values. Water quality and utility characteristics typically lend themselves to such analyses. Six areas of water quality satisfaction (overall quality, taste, smell, appearance, healthiness, and safety) were examined using the following variables: utility capacity (MGD), service population size and percent groundwater used. Average satisfaction levels were computed for each market and then plotted against various utility/water quality parameters to determine the degree of association between two features (Table 5.9). Typically, less than 10% (i.e., R^2 value) of the variation observed in satisfaction levels were explained by water temperature, utility capacity, service population size, or the percent groundwater used. Only two characteristics, TDS and hardness, showed any degree of correlation (R^2 values from 0.34 to 0.64). As expected these water quality parameters were negatively correlated with satisfaction levels; increased levels of hardness or TDS produced lower levels of satisfaction (Fig. 5.6 and 5.7).

Table 5.9
Correlation coefficient (R^2) for variable affection six different water quality features

	Satisfaction with...					
Water Quality/Utility Characteristic	Overall Quality	Taste	Appearance	Smell	Safety	Healthiness
Plant capacity	0.03	0.00	0.09	0.03	0.00	0.01
Service population	0.02	0.00	0.08	0.02	0.01	0.03
% Ground water used	0.02	0.00	0.07	0.04	0.08	0.02
Treated water temp	0.09	0.07	0.04	0.12	0.02	0.10
TDS	0.51	0.37	0.43	0.48	0.36	0.52
Hardness	0.51	0.44	0.42	0.64	0.34	0.37

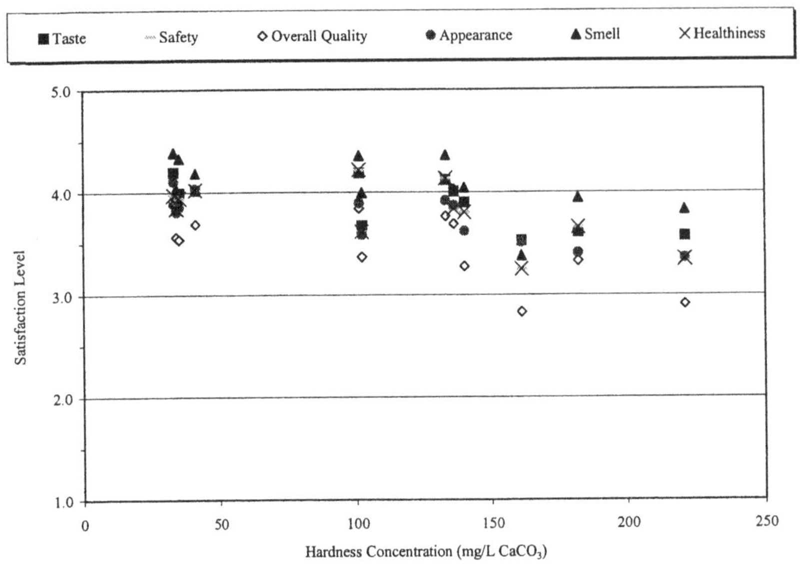

Figure 5.6 Correlation between consumers' satisfaction with overall tap water quality and TDS in tap water

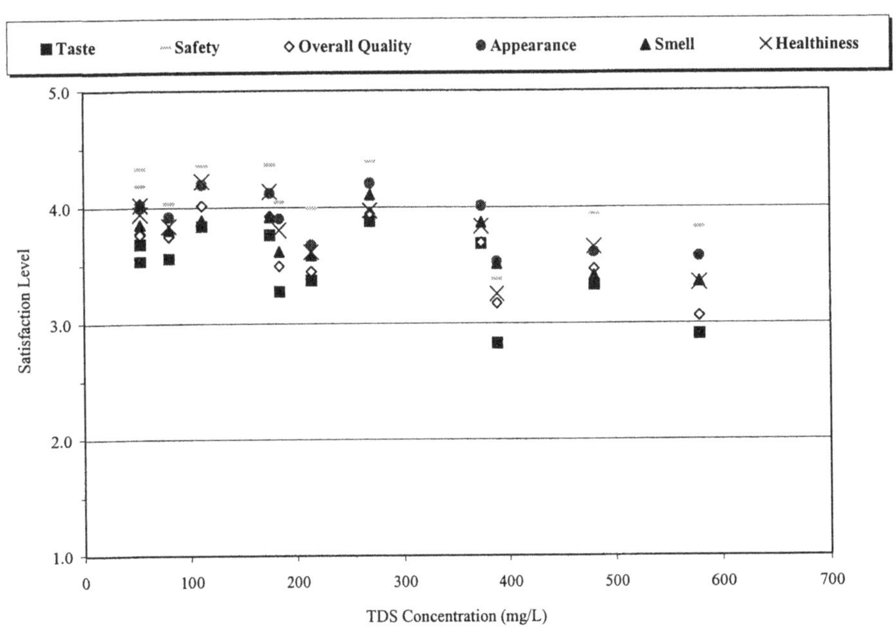

Figure 5.7 Correlation between consumers' satisfaction with overall tap water quality and hardness in tap water

CHAPTER 6
UTILITY MANAGERS' SURVEYS

DEMOGRAPHICS AND BACKGROUND

The Utility Managers' Survey was distributed to water agencies nationwide, covering twenty-two states located in various regions throughout the country. Approximately 130 contacts were made. A total of 117 responses were received from agency representatives associated with 57 utilities (Figure 6.1). A member of upper management was requested to fill out the survey (e.g., plant manager or department head). Some utilities also had "support personnel" (e.g., operators, engineers and customer service) respond.

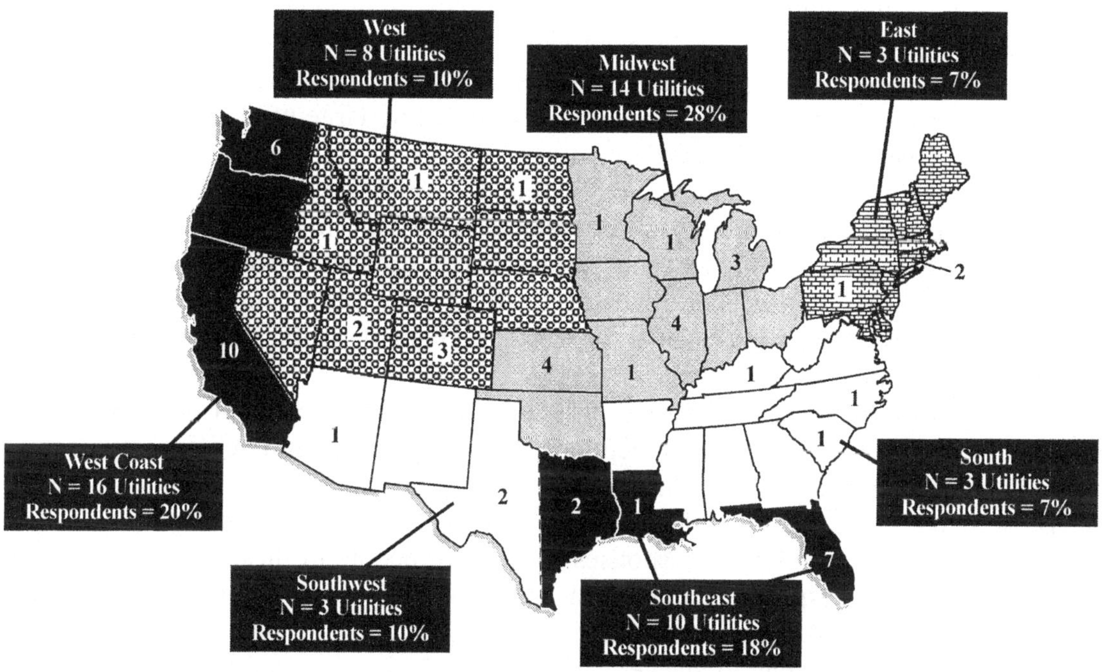

Figure 6.1 Geographic spread of utilities participating in the "utility managers'" survey

The survey was designed to assess utility manager perception of consumers and consumer attitudes toward utilities and to complement the data collected in the customers' survey. While the survey allowed interviewees to respond openly, it was designed to require less than 10 minutes to complete. To promote candid responses, utility managers were informed that the surveys would be published anonymously. A copy of the managers' survey is provided in Appendix E.

The intent behind the "national managers' survey" was to determine in a collective sense the "average utility's perceptions" on these questions. Specifically, it was important to obtain the viewpoints of those in positions that could be most effective at changing management policies. Thus, the larger the sample population the more representative responses will be in terms of a "national average".

In addition, utility managers in the highest level position at a utility are not always in the best position to make a decision and might rely on other division managers (or department heads). For example, the utility manager may not be as familiar with the customer service aspect of operations. Instead, his/her specialty might be operations, contracts/bidding, or office management. In this case, the utility manager would rely on the customer service department head for recommendations regarding customer outreach. Responses from non-management personnel comprise about 1/3 of the total responses.

Figure 6.2 illustrates the breakdown of surveyed utilities by treatment capacity, raw water source, and respondent position. A diverse cross section of utilities were surveyed; the majority of responses came from very large utilities serving 100,000 or more people. Utilities of all sizes, except for those serving more than 1,000,000 people, were represented by at least 10% of the responses. Response diversity based on plant capacity showed a tendency for the larger sized utilities (>20 mgd) to be more represented than smaller utilities. However, this was expected due to the disproportionate number of people in the U.S. served by large utilities; 81% of the U.S. is serviced by 7% of the utilities (EPA 2001b).

Utility diversity based on source water (ground vs. surface water) followed similar trends as reported by the EPA for community water systems in the U.S. In the managers' survey, there was approximately a 60:40 split between surface water user and ground water utilities. An examination of utilities in the U.S. revealed a 70:30 split (EPA 2001b). One measure of diversity, utility ownership, was obviously skewed in favor of public utilities; only 6% of the responses came from private utilities.

The intent of the managers' survey was to ascertain the viewpoint of those in position that could affect consumer attitudes/perceptions. While operators and service engineers could indeed make this happen, typically those in higher, management level positions (e.g., plant managers or department directors) have greater political leverage to make changes happen. Overall, nearly two-thirds (64%) of responses came from people in these managerial positions. Twenty percent of the responses came from "support staff" (e.g., engineers, operators, and secretaries) while the remaining 16% of the respondents failed to state their position on the survey form. Such a refusal rate to answer a question might be expected for sensitive questions. For example, in the consumer survey 20% of the respondents refused to answer the question regarding household income.

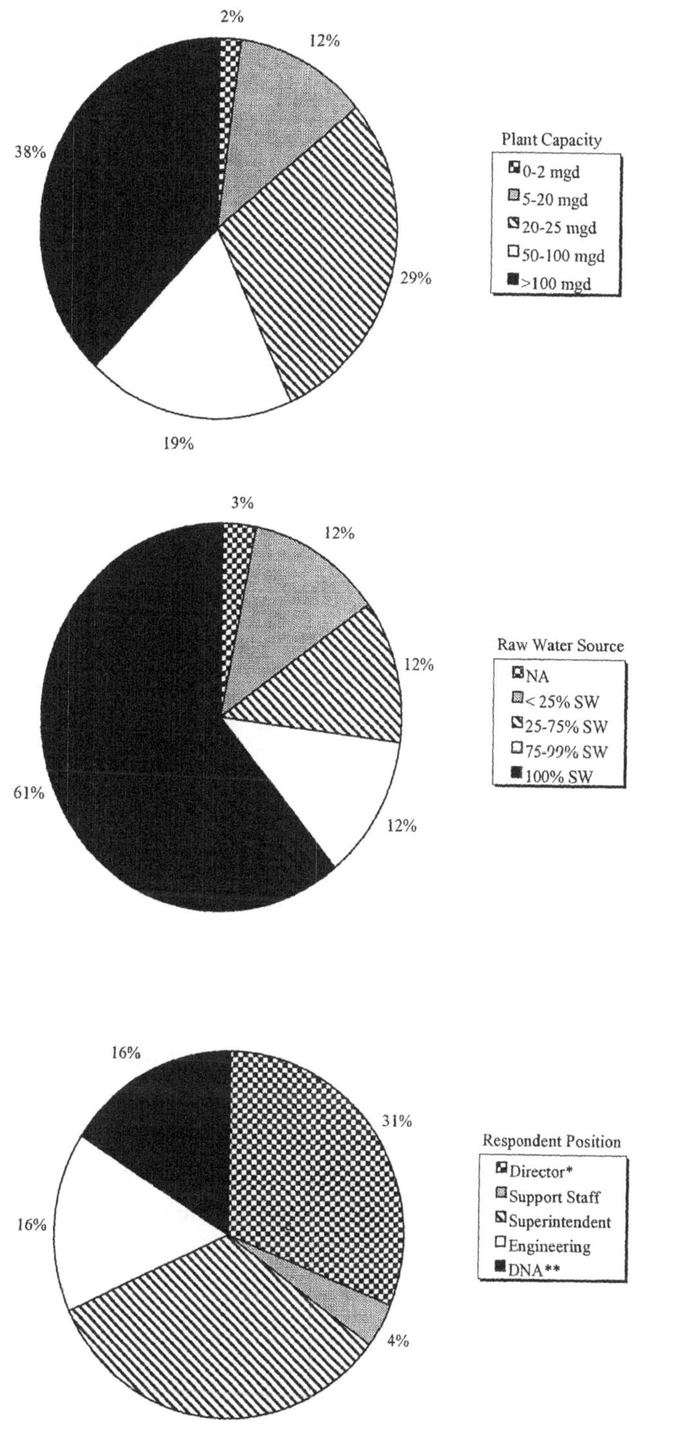

* management heads of specific departments (e.g., water quality)
** DNA = did not answer

Figure 6.2 Classification of surveyed utilities by plant capacity, raw water source and respondent position

CONSUMER SATISFACTION WITH WATER QUALITY

Utility managers were asked to rate their typical customers' satisfaction with the overall quality, taste, safety, and healthiness of their tap water. Their responses are summarized in Table 6.1. Most managers rated their customers as being somewhat satisfied to very satisfied with their tap water quality. They felt that only a small fraction of customers believed that their water was unsafe or unhcalthy and indicated that taste and safety were the two most important water quality parameters to their customers.

Table 6.1
Managers' perceptions of consumer satisfaction with water quality (questions 1, 2, 3, 4 and 5)

	Percent respondents per water quality feature (%)				
Level of satisfaction	Overall quality	Taste	Safety	Most important feature	
Very satisfied	23	15	61	Safety	(27%)
Somewhat satisfied	67	61	28	Smell	(7%)
Neutral	8	20	5	Appearance	(14%)
Somewhat dissatisfied	2	4	5	Taste	(32%)
Very dissatisfied	0	0	0	Healthiness	(8%)
DNA	<1	<1	1		(12%)

DNA did not answer

When response options were consolidated from 5 categories to 3 (dissatisfied, neutral, or satisfied) a consistent trend was observed between utility and consumer responses. Utilities typically overestimated the level of satisfaction of the public and in one case by nearly 20%. Results shown in Figure 6.3 were based on the average responses from the 12 markets ("Consumer Response") and 57 utilities ("Utility Response"). Categories that do not sum to 100% were due to respondents who refused to answer the question. There were no statistical tests used in this section.

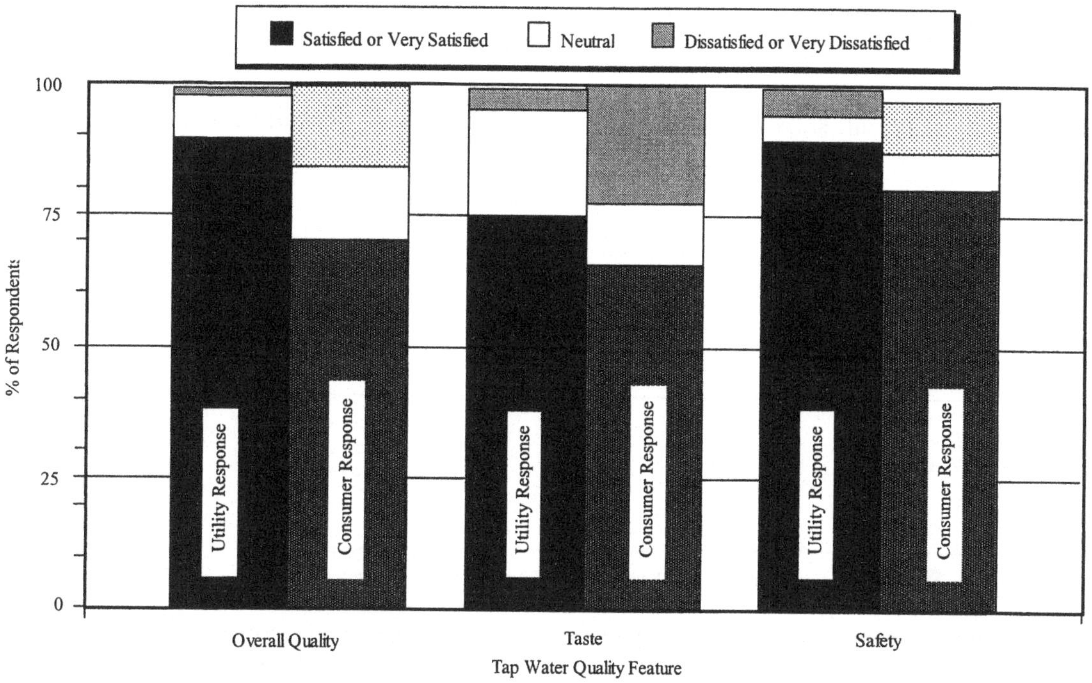

Figure 6.3 Comparison between utility managers and consumers regarding tap water quality features

CONSUMER SATISFACTION WITH LOCAL WATER PROVIDER

When asked to describe their customers' opinions regarding their local water utility, managers suggested that, in general, their customers were satisfied. On average, 80-90% of the managers indicated their customers were satisfied with respect to overall service, customer service, and complaint/problem resolution (Table 6.2). Once again, managers overestimated the level of satisfaction of consumers (Figure 6.4). For all three areas of utility satisfaction, managers overestimated consumer satisfaction by at least 14%. This gap between utility managers' perceptions of consumers' satisfaction level and actual consumer satisfaction levels may, in part, account for the consumption trend toward tap water alternatives.

Table 6.2
Managers' perceptions of consumer satisfaction with local water utilities

Level of satisfaction	Percent respondents per utility attribute		
	Overall satisfaction	Customer service	Resolution of problems or complaints
Very satisficd	16	22	21
Satisfied	72	62	65
Neutral	9	11	9
Very dissatisfied	2	2	3
Dissatisfied	0	1	1
DNA	1	2	1

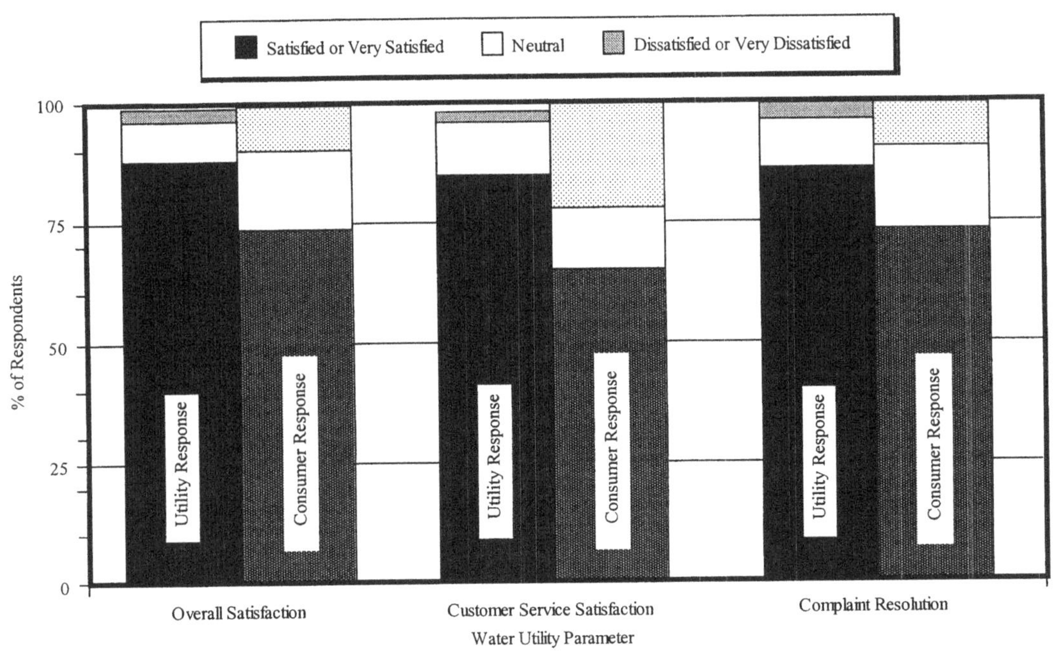

Figure 6.4 Comparison between utility managers and consumers regarding utility satisfaction

Consumer Confidence Reports (CCRs)

The federal requirement to publish annual reports regarding tap water quality and safety has produced mixed reactions from water utilities. In order to better ascertain the true effectiveness of CCRs, all consumers were asked the following question "Within the last year, have you received a Consumer Confidence Report/Annual Drinking Water Quality Report from your local or municipal water provider?" Consumers who did remember receiving a CCR were asked a follow up question: "Did you find this report useful?". The same questions were included in the utility managers' survey to be used for comparison purposes with actual consumer responses.

In spite of the fact that all 12 markets participating in the survey had water utilities that were in compliance with the CCR rule, only about one-third (36%) of the consumers remembered receiving a report. In nine out of the 12 markets, TW drinkers were more likely to remember receiving a CCR than were TWA drinkers (Figure 6.5). Eighty percent of consumers who did remember receiving a CCR found the report useful (Figure 6.6). Utility managers in three of the 12 markets believed the majority of consumers did remember receiving a CCR. Utility managers in eight of the 12 markets believed that the CCR was beneficial to those who at least remembered receiving one.

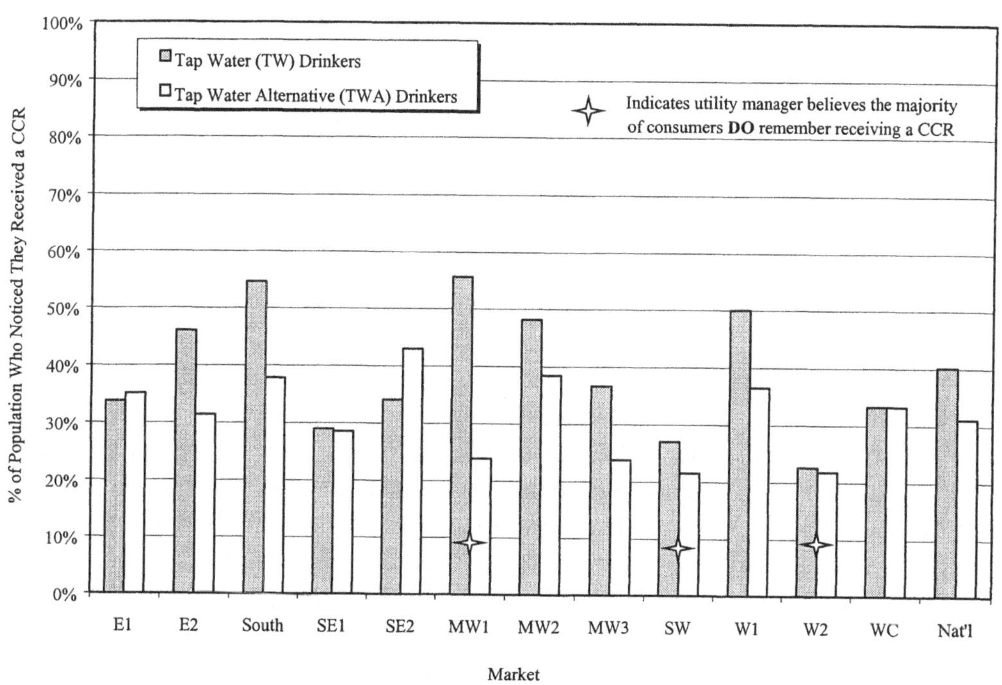

Figure 6.5 Consumers who remembered receiving a CCR/Annual Water Quality Report (Q150)

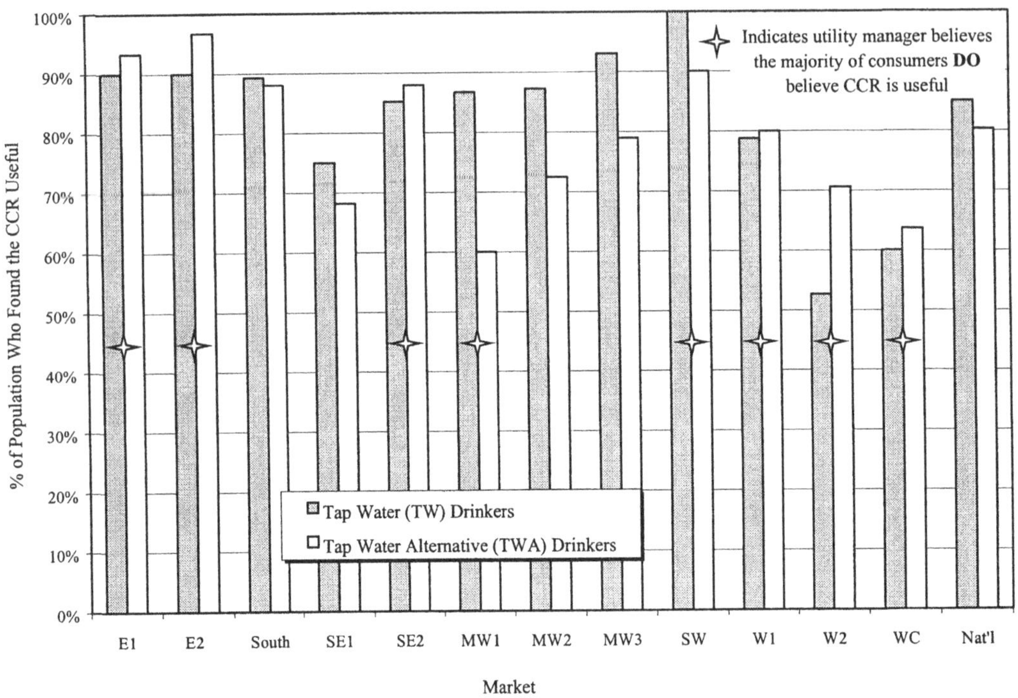

Figure 6.6 Consumers who remembered receiving a CCR <u>and</u> found the report useful (Q155)

Approximately three-quarters of respondents in the national managers' survey said that they have pursued other avenues of customer education besides CCRs. Sixty-four percent of the respondents believed that there are better ways to reach the customer than the current use of CCRs. Some of the customer education tools cited in the managers' survey included the use of newsletters, newspaper articles, educational materials for schools, and water calendar contests.

Motivators for Bottled/Filtered Water Use

According to combined survey results, people who primarily drink bottled water do so for taste, safety and healthiness reasons (approximately 30% each - Figure 6.7). On the other hand, 60% of filtered water drinkers were driven to choose filtered water for safety reasons. The category of "other" in Figure 6.7 includes healthiness, because it was included on the consumer survey but not on the managers' survey. From the managers' perspective, taste was the primary motivator for both bottled and filtered water drinkers (Figures 6.8 and 6.9). In general, managers were overly focused on "taste", and aesthetic issues, rather than fundamental concerns of tap water safety.

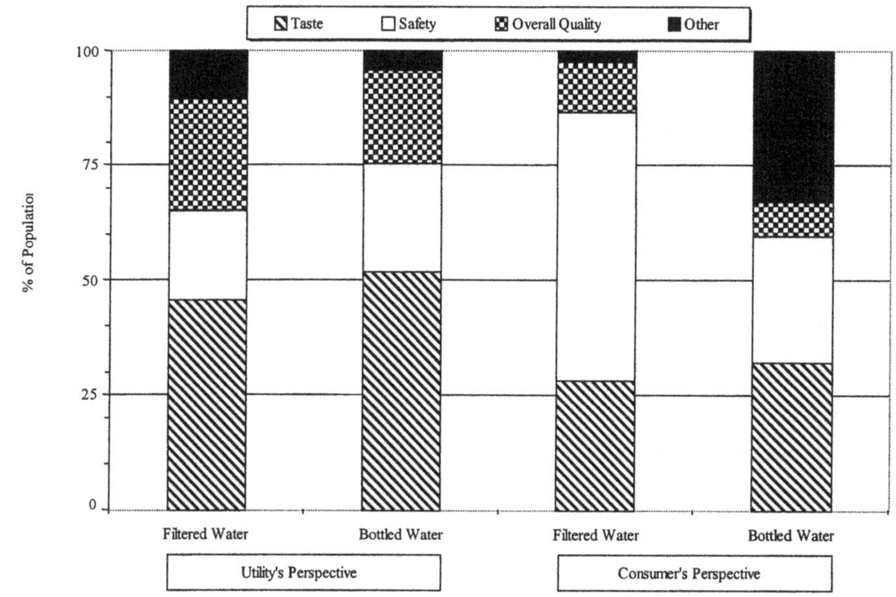

Figure 6.7 Perceived and actual motivators behind bottled and filtered water use

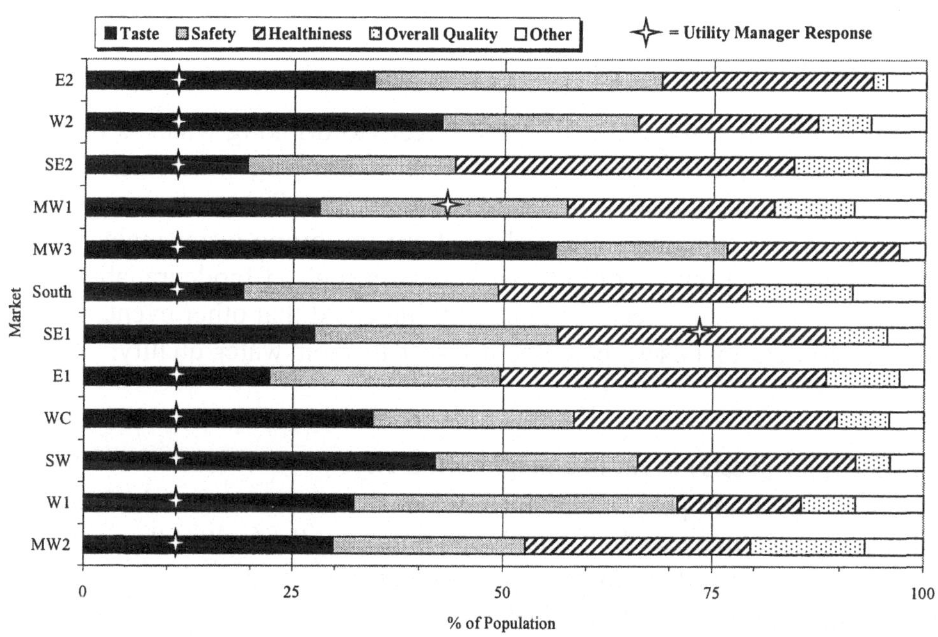

Figure 6.8 Market comparison between the utility and consumers regarding primary motivators for bottled water use (Q102)

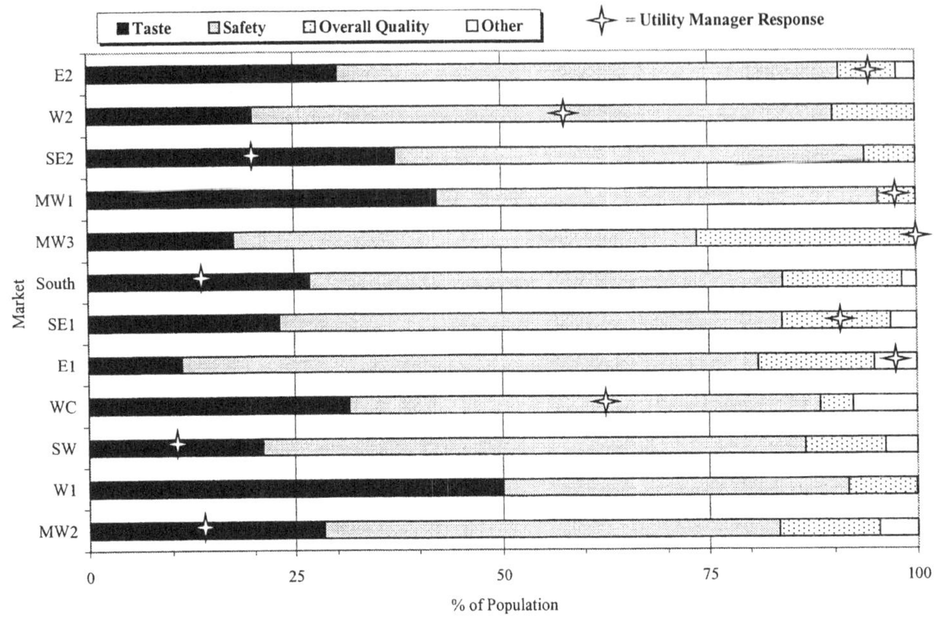

Figure 6.9 Market comparison between the utility and consumers regarding primary motivators for use of a filtration device (Q091)

Perceived/Actual Local Threats to Water Quality

Differentiating between perceived and actual threats to drinking water is often difficult, particularly when the media gets involved. Real threats to drinking water include high TDS breakthroughs (sulfates, chlorides), manganese staining, sewer overflows into raw water sources, main breaks, *Escherichia coli* contamination, sediment stir-up during water main flushing, loss of water pressure during back flow conditions, and breakthrough of biodegradable carbon into distribution systems. Thirty-seven percent of managers indicated that other events occurred that might have caused customers to falsely perceive a threat to their water quality. These events included:

- A taste and odor (T&O) event in a neighboring city
- Events caused by improperly maintained personal water treatment devices, or internal building problems (this may be a real threat, but is not the fault of the utility)
- Detection of rodents in distribution pipes of a neighboring city
- Release of the movie "Erin Brokovich"
- Mishandling of water-related stories by the media
- Diesel spill in a neighboring city
- Toluene spill in a neighboring city that produced metallic taste
- Terrorist attacks of September 11th, 2001
- Publication of general water-related stories in the media (e.g., MTBE, gray water, pesticides, endocrine disruptors)

While nearly half (45%) of consumers remembered a national event, most could only give a vague description of the event. Some typical comments included: "I saw a news show that talked about the quality of the water and sometimes I see it in the newspaper," "something that happened on the East Coast," "New Jersey or something". Arsenic, sewage and chemical spills were often cited. In many cases, the events described were associated with natural disasters (floods) or industrial accidents (chemical spills) and not with drinking water/drinking water treatment directly. Terrorism is also now a concern. A number of respondents polled after the terrorist attacks on the World Trade Center and the Pentagon on September 11, 2001, cited 9/11 as a national event that threatened water quality. A full list of responses by market can be found in the Excel® file "Open ended responses.xls" on the Appendix F CD-ROM.

CHAPTER 7
CONCLUSIONS

SUMMARY

This research has shown that in general, the level of consumer satisfaction with tap water is relatively high in the U.S. As might be expected, people who primarily drink tap water in their household indicated a higher level of satisfaction with tap water than did people who primarily drink a tap water alternative in their household. Taste, healthiness, and safety were areas of greatest dissatisfaction for TWA drinkers and "chlorinous" was the most frequently cited "bad" taste descriptor. When asked to describe their primary reason for choosing to drink a tap water alternative while at home, BW and FW drinkers were driven by safety/health concerns (56% and 68% for BW and FW drinkers respectively) and by taste (30% and 28% respectively). No significant geographic trends were observed with respect to overall tap water quality, taste, odor, safety or healthiness. However, some trends were found when other demographics were related to satisfaction with tap water. In general, consumers under the age of 30 were significantly less satisfied than consumers 65 and older. This is not surprising given that human sensitivity to taste and odors often diminishes with age and that the most significant driver for consumer dissatisfaction was taste. In some markets, households with children were significantly less satisfied with their tap water than were households without children. Differences in satisfaction among consumers based on gender and educational differences were generally not significant. TDS concentration and hardness were the only water quality parameters tested that consistently correlated (negatively) to changes in satisfaction.

Responses to the survey also revealed that while CCRs provide very useful information, the majority of consumers typically are not aware that they receive them. Also, it was observed that some gaps exist between utility managers' perceptions of their consumers' opinions and actual consumer opinions. In order for all parties to make decisions based on real rather than perceived issues, the communication gap between utilities and consumers must be closed.

FUTURIST ANALYSIS

This research has focused on the satisfaction, choices, motivating factors, and knowledge level of today's drinking water consumer. Given the dynamic nature of society, however, this work must be incorporated into a futuristic context. The current U.S. population of 286 million people is expected to reach 350 million people by 2050 (Haaga et al. 2001), which will bring with it an increase in demand for potable water. As technological advances and information availability also expand, people will likely become more discriminate in their choice of products. Consumers will place increasingly higher values on the quality, reliability, and service they receive. One value that has gained significant support in recent years is the focus on health-related products and services, particularly in the food and water industry. The historical goal of preventing acute illness will continue to shift toward a goal of sustaining health over the long term (i.e., chronic illness). For example, the introduction of chlorine as a disinfectant in the water treatment process early in the 20th century reduced the number of microbially induced sicknesses in the U.S from about 20 per 100,000 people per year to less than 1 (Hall 2000). More recent research has shown that disinfection by-products may induce chronic disease such as cancer. In response, some utilities have had to seek alternatives to chlorine to meet disinfection goals. However, "chlorinous" taste and odor descriptors were often used by consumers in this survey when talking about their tap water. Together with the increase in the number of discriminating consumers, the growing emphasis on prevention of chronic illness may ultimately drive the majority of consumers away from tap water consumption.

The expansion of the drinking water market to include bottled water and POU/POE devices also must be considered by utilities. Consumers are being bombarded with an ever-growing selection of drinking water choices. Consequently, new and innovative marketing tactics are continually being developed and used to gain an edge over the competition. For example, many companies advertise that since the resources used to make their product are renewable, they are an environmentally friendly company. Though water utilities have also developed some environmentally conscious programs (e.g. aquifer recharge, wastewater reuse, and low-flow water product development), few consumers are likely to be aware of them. Again, improved communication (i.e., marketing) must be one component in the overall strategy of drinking water utilities.

One final trend that could be considered is increased utility privatization. In general, privatization reduces the financial burden on state and federal governments and can enlarge the pool of funds available for drinking water treatment. However, since companies typically operate with a bottom line involving profit, there is some concern that privatization may come at the expense of customer service. Conversion to private ownership and operation must result in improved water quality and improved consumer education and service.

There are numerous factors that utilities should consider as they develop strategies to compete in the drinking water industry during the next century. It is evident, however, that utilities should start by establishing a good communications program that, at a minimum, conveys the fact that utility water is a sustainable resource, is safe and healthy, and is inexpensive. After all, bottled water distribution generates environmental concerns (e.g. container disposal, emissions from product transportation) and is relatively expensive (~ $1 per gallon compared to $0.001 per gallon for tap water). Another strategy that utilities might consider is the marketing of their water in portable plastic bottles. Though this strategy would diminish a utility's environmental advantage over bottled water companies, it might accomplish two objectives: 1) help utilities compete with the "convenience" factor of bottled water (i.e., being able to carry water anywhere in a resealable container), and 2) serve as a means of communication (product labels could be an effective outreach medium).

RECOMMENDATIONS

Though the quality of the nation's public drinking water supply has experienced significant improvements in recent years, the tap water alternatives market continues to grow at an astounding rate. Given the fact that tap water regulations are stricter than bottled or filtered water regulations, there is no reason for consumers to doubt the safety of public drinking water. However, 58% of FW drinkers claimed that safety is the main reason they use POU/POE devices. BW drinkers also cited safety as one of the primary reasons they use bottled water. Couple these findings with the fact that only two of the 12 utility managers correctly identified FW drinkers' primary concern, and it is clear that there are measures that utilities can take to improve consumer satisfaction in their product.

Satisfaction can be improved by changing water quality (taste, odor, and level of contaminants) and/or by changing the knowledge/perceptions of consumers. A number of recommendations for improving customer satisfaction either through process changes or management changes are put forth in the following two sections. Some of these recommendations came from utility managers and from a review of the literature.

Improving Water Quality

Process changes can be expensive and time consuming. However, sporadic events that only require occasional process modifications should be considered. Elimination of these events would likely have a beneficial impact on customer perception of tap water quality. One particular area that has received a lot of attention recently is taste and odor events. While an auxiliary process may be implemented (e.g., PAC addition) to mitigate taste and odor problems, other areas (e.g., high water temperature) may have no feasible solution. In all cases, it is important to combine any significant process change with public education. With some changes, the water quality might be improved, but the customer may not be aware of it (e.g., implementation of multiple disinfection barriers). In other situations, customers may complain in spite of the fact that water quality is better simply because it is different. The consumer's willingness to pay for an added expense must also be considered. Consumers are typically much more willing to pay for treatment process/distribution system improvements if they are aware of the direct benefit to them. Some examples of issues raised by consumers in the survey where process changes might be applicable are microbial contaminants, microcontaminants, and aesthetics issues.

Microbial Contaminants

- Implement a multiple disinfection barrier strategy to improve/ensure microbial inactivation, (e.g., chlorination/UV, chlorination/ozonation, membrane filtration/UV)

Organic/Inorganic Contaminants

- Implement advanced treatment processes such as granular activated carbon, advanced oxidation, or high-pressure membrane filtration, to remove microcontaminants (e.g., pesticides, endocrine disruptors, heavy metals, etc.).

Aesthetic Issues

- Reduce the concentrations of TDS/hardness in finished water using specific treatment technologies (e.g., lime softening, RO)
- Lower the chlorine residual or switch to chloramines (however switching to chloramines can have other consequences, such as nitrification within the distribution system and so should be carefully considered before being implemented)
- Combat seasonal taste and odor events before they become a significant problem.

Changing Public Perception

While water utilities still operate in a monopolistic market, they must now compete for consumer perception just like in an open market situation. If everyone in the U.S. switched from tap water to bottled/filtered water, the immediate financial repercussions would be trivial. Less than 5% of all water produced by utilities is actually consumed. Most of it is used in watering lawns, washing or industrial applications (AWWA 2001). However, public image and trust is at stake. In an effort to strengthen public confidence, the federal government has mandated the use of CCRS or annual water quality reports. While results in this study showed that CCRs were useful to about 25% of the public, the remaining people must be reached by other methods. People in the U.S. come from diverse social, economic, and cultural backgrounds and as such, no single method can effectively reach them all. Tap, bottled, and filtered water are all basically the same. Differences have been brought to the publics' attention through intense marketing campaigns and it will be marketing that changes the public's current perception. Advertising and "selling" tap water may be a new concept, but to remain in good standing and trust, these new approaches need to be considered by utilities.

Marketing specialists attribute as much as 73% of a successful sale to non-verbal methodologies—i.e., advertisements, attractive logo designs, concern for customer satisfaction (Gallagher 2000). Marketing examples collected in the managers' survey and found in the literature range from simple participation in community events to more elaborate "media teams" organized within the utility (Mackey et al. 2001). Water utilities provide a service and product to their customers. By realigning their marketing efforts with consumer needs, utilities will be well positioned to reasonably provide what the public demands. Some examples of customer education programs collected in this study are listed below:

Special Public Relations (PR) Team

- Organize a PR team and response plan that can deal with negative "events," (e.g., boil water notice, negative media report, or chemical spill in a nearby area). Combat negative press with positive press.
    - It is realized that a utility can only provide information; the utility cannot control how the information is presented. Anecdotal evidence suggests that some media outlets may put a negative "spin" on what a utility presents as positive information to get a larger share of the audience or to increase sales, creating a negative image where a positive one was intended. Methods for developing a proactive, positive news campaign are outside the scope of expertise of the authors. Involvement of public relations experts would likely be well advised before engaging in this type of activity.

Current PR Programs

- Educate the public about source water protection and drinking water treatment. This can be accomplished through tours of the water treatment plant (though heightened safety concerns may make this impossible in some locations), public outreach in local schools, activist groups, and community forums/meetings.
    - Consider an annual open house event for the public at the local water utility

- Participate in local events (e.g., science fairs, county fairs, specialty events)

- Modify CCRs / Annual Water Quality Reports
  - Focus on a marketing approach when designing the report rather than a mass transfer of data
  - Differentiate the Water Quality Report from the water bill and from solicitations. It can easily get tossed out as "junk mail" when combined with a bill. Many companies send out a lot of sales material with their bills. If sent separately however, care should be taken to ensure it is not misidentified as a "junk mail" solicitation.
  - Advertise or publish them in the local paper, on television, billboards or the radio to increase customer awareness
  - Include a FAQ section to address common customer complaints
  - Address local issues related to the water industry (e.g., terrorist threats, infrastructure needs, seasonal water quality problems)
  - Consider using a "Coming Soon" flyer mailed out before the CCR
  - Explain CCR terminology in layman's terms

Alternative Marketing Tactics

- Borrow marketing ideas from private industry techniques similar to bottled and filtered water providers in the "beverage" marketplace. Forty-five percent of respondents in the utility manager's survey responded positively about this option.
  - Develop an advertising campaign.
  - Develop a bottled water program for sale at local stores or for distribution at charity events and related local activities. Get feedback from other utilities that have already developed such programs as you develop your program.
  - Develop a POU/POE device service as an "added value" option for customers. Issues such as potential liability, maintenance, customer acceptance should be investigated before such a costly effort is undertaken.
  - Partner with bottled and/or filtered water companies to provide bottled water and/or filters as an "added value" option for customers.
- Use alternative media besides CCR brochure.
  - Develop a web site with an online CCR and FAQ page.
  - Provide a video showing the overview of water treatment from source to tap.
  - Consider using television, radio or billboards as an advertising tool.

APPENDIX A
CONSUMER SURVEY INSTRUMENT

Note: Text in questions in CAPITALS is prompted on the CATI screen, but not read aloud to the survey respondents.

IntroQ

Hello, my name is <Your Name> calling from Clearwater Research. We are conducting a nationwide study of consumer's perceptions of bottled water and water filtration devices in the United States. This research project seeks to better understand how Americans make decisions about their household drinking water. Your participation is important to the success of this survey.

This telephone number was chosen at random to be included in the survey.

Is this <repeat phone number >?

1. CORRECT NUMBER **(GO TO PRIVRES)**
2. NO ANSWER
3. NORMAL BUSY
4. ANSWERING MACHINE **(LEAVE MESSAGE USE SCRIPT ONE TIME ONLY)**
5. DO NOT WISH TO DIAL THIS NUMBER **(NULL ATTEMPT)**
6. NUMBER IS NOT THE SAME

SHOW ANSWER MACHINE SCRIPT:
Hello, my name is <Your Name> calling from Clearwater Research. We are conducting a nationwide survey of consumer's perceptions of bottled water and water filtration devices in the United States. This telephone number was chosen at random to be included in the survey. I would like to speak with an adult in your household and will be calling back within a couple of days. Your participation is important to the success of this survey. Thank you.

PrivRes

Is this a private <u>residence</u>?

1. YES, CONTINUE
2. NO, NON-RESIDENTIAL

NonRes - **Only get this if PrivRes = 2** (Non-Residential)

Thank you very much, but we are only interviewing private residences.

WrongNum - **ONLY GET THIS IF INTROQ = 6 (NUMBER IS NOT THE SAME)**

Thank you very much. I seem to have dialed the wrong number. It's possible that your number may be called at a later time.

Adults1 **EVERYONE GETS**

I would like to speak with a person 18 years or older only to ask some questions regarding your drinking water. May I speak with an adult 18 years or older please?

1. YES, I AM AN ADULT.
2. YES, ADULT COMING TO THE PHONE.
3. NO, GO TO THE NEXT SCREEN, PRESS CTRL END AND SCHEDULE A CALL-BACK

*** DO NOT USE CTRL-END ON THIS SCREEN ***

Adults2 - **ONLY IF ADULTS1 = 1 OR ADULTS1=2 (YES) OR YOURTHE1 = 2**

Our survey requires that we randomly select one adult who lives in your household to be interviewed. How many members of your household, including yourself, are 18 years of age or older?

__ ENTER THE NUMBER OF ADULTS

Men **ONLY GET IF ADULTS>1**

How many of these adults are men?

1. None
2. One
3. Two
4. Three

5. Four
6. Five
7. Six
8. Seven
9. Eight
10. Nine

Women **ONLY GET IF ADULTS>1 AND MEN<ADULTS**

How many of these adults are women?

1. None
2. One
3. Two
4. Three
5. Four
6. Five
7. Six
8. Seven
9. Eight
10. Nine

WrongTot **ONLY GET IF ADULTS>1 AND MEN+WOMEN<>ADULTS**

Probe: I'm sorry, something is not right.

Number of men -
Number of women - +

Number of Adults-

1. Correct the number of men **GO TO MEN**
2. Correct the number of women **GO TO WOMEN**
3. Correct the number of adults **GO TO ADULTS**

Selected – **ONLY GET THIS IF MORE THAN ONE ADULT IN HOUSEHOLD**

The person in your household I need to speak with is the ____________.

Are you the ____________?

1. YES **GO TO YOURTHE1**
2. NO **GO TO GETADULT**

GetAdult –**ONLY GET THIS IF SELECTED=2**

May I speak with him or her?

1. YES, SELECTED ADULT COMING TO PHONE
2. NO, GOT TO THE NEXT SCREEN, PRESS CTRL END AND SCHEDULE A CALL-BACK

*** DO NOT USE CTRL-END ON THIS SCREEN ***

Yourthe1 - **ONLY IF ADULTS1 = 1 AND ADULTS2 = 1 OR IF SELECTED = 1 (YES) OR GETADULT = 1**

Then you are the person I need to speak with. Your participation is voluntary, but it is very important because you represent many other people in your community. We will hold your responses in the strictest confidence, but you may decline to answer any question you wish. If you have any questions at any point, please let me know.

This survey covers a wide range of topics concerning bottled and household water. It will take approximately 10 minutes, but may take longer based on your answers. I'll read as quickly as I can.

1. PERSON INTERESTED, CONTINUE
2. PERSON NOT INTERESTED, CLOSING2

PrescrnQ1

Which of the following best describes how you receive your household water?

Would you say your water comes from . . .

1. A municipal or public water company **(GOTO PRESCRNQ4)**
2. A private water company**(GOTO PRESCRNQ4)**
3. A private well on your property**(TERMINATE DISP 13)**
4. OTHER:_____________**(GOTO PRESCRNQ2)**
7. DON'T KNOW / I DO NOT UNDERSTAND**(GOTO PRESCRNQ2)**
9. REFUSED**(TERMINATE DISP 12)**

PrescrnQ2 **(GET IF PRESCRNQ1=4 or PRESCRNQ1=7)**

Do you receive a monthly or quarterly bill for your household water?

1. YES **(GOTO PRESCRNQ1, READ OPTIONS 1 & 2 AGAIN)**
2. NO **(GOTO PRESCRNQ3)**
7. DON'T KNOW / I DO NOT UNDERSTAND**(TERMINATE DISP 12)**
9. REFUSED**(TERMINATE DISP 12)**

PrescrnQ3 **(GET ONLY IF PRESCRNQ1=7)**

Do you currently rent your place of residence?

1. YES**(GOTO PRESCRNQ4)**
2. NO**(TERMINATE DISP 13)**
7. DON'T KNOW / I DO NOT UNDERSTAND**(TERMINATE DISP 12)**
9. REFUSED**(TERMINATE DISP 12)**

PrescrnQ4

Are you currently an employee of a private or municipal water company, bottled water or beverage company, or water treatment manufacturer or retailer?

1. YES**(TERMINATE DISP 14)**
2. NO**(GOTO QUOTAQ)**
7. DON'T KNOW / I DO NOT UNDERSTAND**(TERMINATE DISP 12)**
9. REFUSED**(TERMINATE DISP 12)**

QuotaQ

Which of the following best describes the way you most often drink water in your household.

Would you say you usually drink. . .

1. Tap water straight from the faucet (*i.e.*, unfiltered)**QUOTA CELL A**
2. Tap water that has been filtered or treated within your household**QUOTA CELL B**
3. Bottled water**QUOTA CELL B**
7. DON'T KNOW / I DO NOT UNDERSTAND **TERMINATE DISP 12**
9. REFUSED **TERMINATE DISP 12**

[**NOTE: IF QUOTA CELL IS FULL, THEN READ FOLLOWING SCRIPT:** Thank you for your interest in this research project. We already have enough respondents with that particular drinking preference. Thank you for your time. **TERMINATE: DISPOSITION DISP 11**]

FIRST Section *Perception of Water Quality*

The following set of questions asks about your personal satisfaction with the quality of the tap water you receive in your household.

Let me stress again there are no correct answers. Just pick the answer that best reflects your response to the question.

Q005

Which of the following best describes your satisfaction with the overall quality of the tap water you receive in your household?

Would you say you were. . .

1. Very Dissatisfied
2. Dissatisfied
3. Neither Dissatisfied nor Satisfied
4. Satisfied
5. Very Satisfied
7. DON'T KNOW / I DO NOT UNDERSTAND
9. REFUSED

Q010

Which of the following best describes your satisfaction with the taste of the tap water you receive in your household?

Would you say you were. . .

1. Very Dissatisfied
2. Dissatisfied
3. Neither Dissatisfied nor Satisfied
4. Satisfied
5. Very Satisfied
7. DON'T KNOW / I DO NOT UNDERSTAND
9. REFUSED

Q015

Which of the following best describes your satisfaction with the appearance or look of the tap water you receive in your household?

Would you say you were. . .

1. Very Dissatisfied
2. Dissatisfied
3. Neither Dissatisfied nor Satisfied
4. Satisfied
5. Very Satisfied
7. DON'T KNOW / I DO NOT UNDERSTAND
9. REFUSED

Q020

Which of the following best describes your satisfaction with the smell of the tap water you receive in your household?

Would you say you were. . .

1. Very Dissatisfied
2. Dissatisfied
3. Neither Dissatisfied nor Satisfied
4. Satisfied
5. Very Satisfied
7. DON'T KNOW / I DO NOT UNDERSTAND
9. REFUSED

Q025

Which of the following best describes how safe the tap water you receive in your household is to drink?

Would you say your water is . . .

1. Unsafe
2. Somewhat unsafe
3. Neither unsafe nor safe
4. Somewhat safe
5. Safe

7. DON'T KNOW / I DO NOT UNDERSTAND
9. REFUSED

Q026

Which of the following best describes how healthy the tap water you receive in your household is to drink?

Would you say your water is . . .

1. Unhealthy
2. Somewhat unhealthy
3. Neither unhealthy nor healthy
4. Somewhat healthy
5. Healthy
7. DON'T KNOW / I DO NOT UNDERSTAND
9. REFUSED

Q030

Considering all of these water quality features, which one is the most important to you?

Would you say it was your water's. . .

1. Taste
2. Appearance
3. Smell
4. Safety
5. Healthfulness
7. DON'T KNOW / I DO NOT UNDERSTAND
9. REFUSED

(RANDOMIZE LIST FOR PRESENTATION)

Q035

Do you drink the tap water you receive in your household?

1. YES
2. NO — **SKIP TO Q042**
7. DON'T KNOW / I DO NOT UNDERSTAND
9. REFUSED

Q040

Approximately how many glasses of tap water do you personally drink during a full day? (I.E., AT WORK AND HOME)

1. _______
77. DON'T KNOW / I DO NOT UNDERSTAND
99. REFUSED

Q042

Approximately how many glasses of bottled or filtered water do you personally drink during a full day? (I.E., AT WORK AND HOME)

1. _______
77. DON'T KNOW / I DO NOT UNDERSTAND
99. REFUSED

Q045

How would you describe the taste of the tap water you receive in your household?

11. NONE/NOTHING
12. STALE/FLAT
13. CHLORINOUS (MEDICINAL, SWIMMING POOL)
14. MUSTY/EARTHY
15. SALTY (SALINE)
16. METALLIC
17. PLASTIC
18. FRESH
19. OTHER:_______________
77. DON'T KNOW / I DO NOT UNDERSTAND
88. FINE / OK
99. REFUSED

Q046 **(GET IF Q045=12-17, 19)**

Is there a specific time of year when these tastes are apparent?

1. YES
2. NO
7. DON'T KNOW/DON'T UNDERSTAND
9. REFUSED

Q050

How would you describe the smell of the tap water you receive in your household?

1. NONE – **SKIP TO Q055**
2. CHLORINOUS
3. MUSTY/EARTHY
4. PLASTIC
8. OTHER:______________
7. DON'T KNOW / I DO NOT UNDERSTAND
9. REFUSED

Q051

How often does your water smell <INSERT RESPONSE Q050 HERE>?

1. DAILY
2. AT LEAST ONCE A WEEK
3. AT LEAST ONCE A MONTH
4. AT LEAST ONCE A YEAR
5. ONCE IN THE PAST FIVE YEARS OR LESS
7. DON'T KNOW / I DO NOT UNDERSTAND
8. REFUSED

Q052 **(GET IF Q045=12-17, 19)**

Is there a specific time of day when these odors are apparent?

1 YES
2 NO
3 NA
7 DON'T KNOW / I DO NOT UNDERSTAND
9 REFUSED

Q053 **(GET IF Q055=2-4)**

Is there a specific time of year when these odors are apparent?

1 YES
2. NO
7. DON'T KNOW / DON'T UNDERSTAND
9. REFUSED

Q055

How would you describe the appearance or look of the tap water you receive in your household?

1. CLEAR
2. COLORED — **GO TO Q060**
3. CLOUDY (TURBID, UNCLEAR)
4. OTHER:_____________
7. DON'T KNOW / I DO NOT UNDERSTAND
9. REFUSED

Q060 **GET IF Q055 = 2**

What color?

1. RED
2. ORANGE
3. YELLOW
4. BROWN
5. BLACK
6. WHITE
8. OTHER:_____________
7. DON'T KNOW / I DO NOT UNDERSTAND
9. REFUSED

Q061 **GET IF Q055=1, 7, or 9**

How often would you say that your water looks <INSERT RESPONSE Q055 HERE>?

1. DAILY

2. AT LEAST ONCE A WEEK
3. AT LEAST ONCE A MONTH
4. AT LEAST ONCE A YEAR
5. ONCE IN THE PAST FIVE YEARS OR LESS
7. DON'T KNOW / I DO NOT UNDERSTAND
9. REFUSED

Q063

Do you use a water-softening unit for the tap water you receive in your household?

1. YES
2. NO
7. DON'T KNOW / I DO NOT UNDERSTAND
9. REFUSED

Q064 **GET IF Q063 = 1**

Does this water softening unit treat only the hot water, only the water coming out of your tap, or both?

1. ONLY THE HOT WATER
2. ONLY AT THE TAP
3. BOTH (ALL THE TAP WATER)
7. DON'T KNOW / I DO NOT UNDERSTAND
9. REFUSED

Q065 **GET IF Q063=1**

Other than a water softener, do you use a water filtration system for the tap water you receive in your household?

1. YES
2. NO — **SKIP TO Q095**
7. DON'T KNOW / I DO NOT UNDERSTAND
9. REFUSED

Q070 **GET IF Q065 = 1**

How many filtration systems do you have installed in your household?

1. ______
7. DON'T KNOW / I DO NOT UNDERSTAND
9. REFUSED

Q080 **GET IF Q065 = 1**

Where is your filtration system installed?

1. _________
7. DON'T KNOW / I DO NOT UNDERSTAND
9. REFUSED

(IF Q070 > 1, THEN REPEAT Q080 FOR ALL N)

Q085 **GET IF Q065 = 1**

How long has your filtration system been installed in your household?

1. _________
7. DON'T KNOW / I DO NOT UNDERSTAND
9. REFUSED

Q090 **GET IF Q065 = 1**

Which of the following water quality issues are reasons you use a water filtration system?

1. Taste
2. Appearance
3. Smell
4. Safety
5. Overall quality
6. OTHER:__________
7. DON'T KNOW / I DO NOT UNDERSTAND
9. REFUSED

(RANDOMIZE LIST FOR PRESENTATION)

Q091 **GET IF Q065 = 1**

Of these issues, which is the most important? Would you say it was. . .

3. Taste
4. Appearance
5. Smell
6. Safety
7. Overall quality
8. OTHER:__________
7. DON'T KNOW / I DO NOT UNDERSTAND
9. REFUSED

(RANDOMIZE LIST FOR PRESENTATION)

Q092 **GET IF Q065 = 1**

Approximately how much money do you spend every month on maintaining your water filtration system?

1. __________
7. DON'T KNOW / I DO NOT UNDERSTAND
9. REFUSED

Q095

Do you drink bottled water while at home?

1. YES
2. NO — **SKIP TO Q110**
7. DON'T KNOW / I DO NOT UNDERSTAND
9. REFUSED

Q100 **GET IF Q095 = 1**

How many glasses of bottled water do you personally drink during a full day? (I.E., WORK AND HOME)

1. ________
7. DON'T KNOW / I DO NOT UNDERSTAND
9. REFUSED

Q101 **GET IF Q095/Q042 = 1**

Which of the following water quality issues are reasons you use bottled water?

(RANDOMIZE LIST FOR PRESENTATION)

1. Taste
2. Appearance
3. Smell
4. Safety
5. Overall quality
6. Healthiness
8. OTHER:__________
7. DON'T KNOW / I DO NOT UNDERSTAND
9. REFUSED

Q102 **GET IF Q095/Q042 = 1**

Of these issues, which is the most important?

1. Taste
2. Appearance
3. Smell
4. Safety
5. Overall quality
6. Healthiness
7. OTHER:__________
7. DON'T KNOW / I DO NOT UNDERSTAND
9. REFUSED

(RANDOMIZE LIST FOR PRESENTATION)

Q105 GET IF Q095/Q042=1

Indicate the categories in which you believe bottled water is better than tap water.

1. Taste
2. Smell
3. Safety
4. Healthiness
5. Quality
6. Cost

(RANDOMIZE LIST FOR PRESENTATION)

SECOND Section *Relation with Local Water Provider*

The following set of questions asks about your opinions regarding your local water provider.

Q110

Which of the following best describes how confident you are in the safety of the tap water you receive in your household?

Would you say you were. . .

1. Not Very Confident
2. Not Confident
3. Neither Not Confident nor Confident
4. Confident
5. Very Confident
7. DON'T KNOW / I DO NOT UNDERSTAND
9. REFUSED

Q115

Which of the following best describes how confident you are that the tap water you receive in your household meets or exceeds state and federal requirements?

Would you say you were. . .

1. Not Very Confident
2. Not Confident
3. Neither Not Confident nor Confident
4. Confident
5. Very Confident
7. DON'T KNOW / I DO NOT UNDERSTAND
9. REFUSED

Q120

Which of the following best describes your overall satisfaction with your local or municipal water provider?

Would you say you were. . .

1. Very Dissatisfied
2. Dissatisfied
3. Neither Dissatisfied nor Satisfied
4. Satisfied
5. Very Satisfied
7. DON'T KNOW / I DO NOT UNDERSTAND
9. REFUSED

Q130

Which of the following best describes your satisfaction with the customer service provided by your local or municipal water provider?

Would you say you were. . .

1. Very Dissatisfied
2. Dissatisfied
3. Neither Dissatisfied nor Satisfied

4. Satisfied
5. Very Satisfied
7. DON'T KNOW / I DO NOT UNDERSTAND
9. REFUSED

Q135

Within the past year, have you personally contacted your local or municipal water provider with a question or complaint?

1. YES
2. NO — **SKIP TO Q145**
7. DON'T KNOW / I DO NOT UNDERSTAND
9. REFUSED

Q140 **GET IF Q135=1**

Which of the following best describes your level of satisfaction with the result of your question or complaint?

Would you say you were. . .

1. Very Dissatisfied
2. Dissatisfied
3. Neither Dissatisfied nor Satisfied
4. Satisfied
5. Very Satisfied
7. DON'T KNOW / I DO NOT UNDERSTAND
9. REFUSED

Q145

Which of the following best describes your level of satisfaction with water quality information provided by your local or municipal water provider?

Would you say you were. . .

1. Very Dissatisfied
2. Dissatisfied
3. Neither Dissatisfied nor Satisfied

4. Satisfied
5. Very Satisfied
7. DON'T KNOW / I DO NOT UNDERSTAND
9. REFUSED

Q150

Within the last year, have you received a Consumer Confidence Report/Annual Drinking Water Quality Report from your local or municipal water provider?

1. YES
2. NO — **SKIP TO Q165**
7. DON'T KNOW / I DO NOT UNDERSTAND
9. REFUSED

Q155

Did you find this report useful?

1. YES
2. NO
7. DON'T KNOW / I DO NOT UNDERSTAND
9. REFUSED

Q165

Are you aware of any consumer education programs initiated by your local or municipal water provider?

1. YES
2. NO
7. DON'T KNOW / I DO NOT UNDERSTAND
9. REFUSED

Q190

Within the last year, are you aware of any local incident that seriously threatened the quality of water you receive in your household?

1. YES
2. NO — **SKIP TO Q200**
7. DON'T KNOW / I DO NOT UNDERSTAND
9. REFUSED

Q195 **GET IF Q190 = 1**

Can you briefly describe the nature of this local incident that seriously threatened the quality of water you receive in your household?

1.___________
7. DON'T KNOW / I DO NOT UNDERSTAND
9. REFUSED

Q200

Within the last year, are you aware of any national incident that seriously threatened the quality of water in some other community?

1. YES
2. NO
7 DON'T KNOW / I DO NOT UNDERSTAND
9. REFUSED

Q205 **GET IF Q200 = 1**

Where did this national incident that seriously threatened the quality of water occur?

1. ___________
7. DON'T KNOW / I DO NOT UNDERSTAND
9. REFUSED

Q210 **GET IF Q200 = 1**

Can you briefly describe the nature of this national incident that seriously threatened the quality of water in some other community?

1. ___________
7. DON'T KNOW / I DO NOT UNDERSTAND
9. REFUSED

THIRD Section *Demographics*

Now, I'm going to ask a set of general questions about your personal characteristics. When I read each one, please answer to the best of your ability.

Q215

When is your birthday?

RECORD: MM/DD/YY

77 = DON'T KNOW / NOT SURE
99 = REFUSED

Q220 **GET IF Q215 = 99**

[Interviewer: Script about confidentiality and importance of respondent's cooperation]

What is your age?

XXX

777 DON'T KNOW / NOT SURE
999 REFUSED

Q225 **GET IF Q220 = 999**

[Interviewer: Stress confidentiality and importance to data validity]

I understand this is a sensitive question. Would you be willing to tell me which of the following best describes your age range?

[READ CHOICES]

1. Under 30
2. 30 to 44
3. 45 to 64

4. 65 or older
777DON'T KNOW / NOT SURE
999 REFUSED

Q230

[INTERVIEWER: RECORD GENDER, VERIFY IF NECESSARY]

1. MALE
2. FEMALE
7. DON'T KNOW / NOT SURE
9. REFUSED

Q235

Which racial or ethnic background best describes you? Would you say: White, Black, Asian, Pacific Islander, American Indian, Alaska Native, or Other?

[READ CHOICES]

1. White or Caucasian
2. Black or African-American
3. Hispanic or Latin American
4. Asian or Pacific Islander
5. Native American or Aleut (Al-Oot)?
6. OTHER (Specify?)
7. DON'T KNOW / NOT SURE
9. REFUSED

Q245

How long have you lived in the apartment or household in which you currently reside?

RECORD: XXX

777 = DON'T KNOW / NOT SURE
999 = REFUSED

Q250

What is the highest grade or year of school you completed?

[READ ONLY IF NECESSARY]

1. NEVER ATTENDED SCHOOL OR ONLY ATTENDED KINDERGARTEN
2. GRADES 1 THROUGH 8 (ELEMENTARY)
3. GRADES 9 THROUGH 11 (SOME HIGH SCHOOL)
4. GRADE 12 OR GED (HIGH SCHOOL GRADUATE)
5. COLLEGE 1 YEAR TO 3 YEARS (SOME COLLEGE OR TECHNICAL SCHOOL)
6. COMPLETION OF A 2-YEAR TECHNICAL OR ASSOCIATE DEGREE
7. COLLEGE 4 YEARS OR MORE (COLLEGE GRADUATE)
8. GRADUATE SCHOOL
9. REFUSED

Q255

How many children aged 17 and below reside in your household:

XX [ENTER AMOUNT]

20 = 20 or more
77 = DON'T KNOW / NOT SURE
99 = REFUSED

Q260 **IF Q255=1**

What is the child's age?

XX [ENTER AMOUNT]

77 = DON'T KNOW / NOT SURE
99 = REFUSED

Q265 **IF Q255>1**

What are the children's ages?

XX [ENTER AMOUNTS, SEPARATED BY COMMAS]

77 = DON'T KNOW / NOT SURE
99 = REFUSED

Q266

Do you smoke or use tobacco products?

1. YES
2. NO
7. DON'T KNOW / I DO NOT UNDERSTAND
9. REFUSED

Q300

Including all possible sources of income, what was the total income for your household last year?

$XXX,XXX,XXX [ENTER AMOUNT]

777777777. DON'T KNOW / I DO NOT UNDERSTAND
999999999. REFUSED

Q305 **GET IF Q300 = 999999999**

[INTERVIEWER: AGAIN STRESS CONFIDENTIALITY AND IMPORTANCE OF COMPLETE DATA]

Including all possible sources of income, which of the following ranges best describes the total income for your household last year?

1. Under $15,000
2. At least $15,000 but no more than $25,000
3. At least $25,000 but no more than $35,000
4. At least $35,000 but no more than $45,000
5. At least $45,000 but no more than $55,000
6. At least $55,000 but no more than $75,000
7. At least $75,000 but no more than $100,000
8. At least $100,000 but no more than $250,000
9. More than $250,000
10. DON'T KNOW / I DO NOT UNDERSTAND
11. REFUSED

CLOSING

Thank you very much for helping us in our study. This research project would not be possible without the help of people like you, and we are deeply grateful for your participation.

APPENDIX B
CATI VARIABLES

CATI Code	Description	Coding algorithm	Step
21	Answering Machine	If one disposition code in the call history was either 03 (Answering Machine) or 04 (Answering Machine Message).	**6**
22	Refusal	If the disposition code in the call history was 12 (Initial Refusal).	**3**
23	Disconnect/Non Working	If all dispositions in the call history were 14 (Fast Busy or Noise).	**7**
24	No Answer	If the majority of dispositions in the call history were 01 (No Answer).	**8**
25	Not a Private Residence	If one disposition code in the call history was 25 (Not a Private Residence).	**11**
26	No Eligible Respondent at Number	If one disposition code in the call history was 26 (No Eligible Respondent at Number).	**12**
27	No Eligible Respondent During Interview Period	If one disposition code in the call history was 11 (Definite Appointment or Callback).	**4**
28	Language Barrier	If one disposition code in the call history was 15 (Non-English Speaking).	**5**
29	Interview Terminated in Questionnaire	If one disposition code in the call history was 13 (Initial Terminate in Questionnaire).	**2**
30	Line Busy	If most of the disposition codes in the call history were 02 (Busy) or 14 (Fast Busy or Noise).	**9**
31	Respondent Unable to Communicate	If one disposition code in the call history was 31 (Respondent Unable to Communicate).	**10**
36	Complete	If final disposition code in the call history was 36 (Complete).	**1**

Note: The assignment of unique CATI codes is based on the stepwise application each of the previously listed algorithms. The specific order for the replacement of last disposition codes with final disposition codes is provided in the step column. Once a final disposition has been assigned, no further changes are made based on subsequent steps.

APPENDIX C
RESPONSE RATE CALCULATION FORMULAS

Note: The following formulas were used with raw data collected from each market to calculate response rate, upper bound and refusal rate. Formula variables are given as numbers. An explanation of these codes is presented in Appendix B.

The following is a brief description of each formula. "Response Rate" is the number of respondents who successfully completely the survey out of all residential numbers dialed. "Upper bound" is a liberal estimate of the "response rate" and essentially eliminates all unknown households as potential respondents. "Refusal rate" is the number of eligible respondents who chose not to complete the survey.

1. Council of American Survey Research Organizations (CASRO) Response Rate Calculation Formula

$$\frac{36}{(36+22+27+29)+\dfrac{(36+22+27+29)}{(36+22+27+29)+(23+25+26+28+31)}\times(21+24+30)}$$

2. Upper Bound

$$\frac{36}{(36+22+29)}$$

3. Refusal Rate

$$\frac{22}{(21+22+23+24+25+26+27+28+29+30+31+36)}$$

APPENDIX D
RESPONSE RATES BY MARKET AREA

1. West Coast

Household	
Refusal (Eligibility Unknown)	659
Complete	123
No Eligible Respondent During Interview	113
Quota Cell Full	41
Final Termination in Questionnaire	11
No Eligible Respondent at Number	1
Total Household	948
Non-Household	
Not a Private Residence	450
Disconnect/ Non-working	342
Total Non-Household	792
Unknown	
Final No Answer	378
Language Barrier	50
Don't Know	10
Final Line Busy	22
Total Unknown	460
Total Dispositions	2200

CASRO	10.26%
Upper Bound	15.51%
Refusal Rate	30.45%

2. Southwest

Household	
No Eligible Respondent at Number	2
Quota Cell Full	182
Final Termination in Questionnaire	26
Refusal (Eligibility Unknown)	1157
Complete	159
No Eligible Respondent During Interview	235
Total Household	1761
Non-Household	
Not a Private Residence	1400
Disconnect/ Non-working	3201
Total Non-Household	4601
Unknown	
Final No Answer	891
Language Barrier	112
Final Line Busy	155
Total Unknown	1158
Total Dispositions	7520

CASRO	7.64%
Upper Bound	11.85%
Refusal Rate	15.73%

3. West 1

Household	
No Eligible Respondent at Number	2
No Eligible Respondent During Interview	208
Final Termination in Questionnaire	12
Refusal (Eligibility Unknown)	690
Complete	200
Quota Cell Full	36
Total Household	1148
Non-Household	
Not a Private Residence	40
Disconnect/ Non-working	76
Total Non-Household	116
Unknown	
Don't Know	1
Final No Answer	196
Language Barrier	15
Final Line Busy	5
Total Unknown	217
Total Dispositions	1481

CASRO	14.87%
Upper Bound	22.17%
Refusal Rate	47.40%

4. West 2

Household	
Refusal (Eligibility Unknown)	998
Complete	200
Final Termination in Questionnaire	13
No Eligible Respondent During Interview	170
No Eligible Respondent at Number	3
Quota Cell Full	84
Total Household	1468
Non-Household	
Not a Private Residence	601
Disconnect/ Non-working	1115
Total Non-Household	1716
Unknown	
Language Barrier	41
Final Line Busy	29
Final No Answer	546
Total Unknown	616
Total Dispositions	3800

CASRO	11.42%
Upper Bound	16.52%
Refusal Rate	26.61%

5. Midwest 1

Household	
Final Termination in Questionnaire	5
No Eligible Respondent at Number	3
Complete	200
Quota Cell Full	37
No Eligible Respondent During Interview	212
Refusal (Eligibility Unknown)	993
Total Household	1450
Non-Household	
Disconnect/ Non-working	964
Not a Private Residence	748
Total Non-Household	1712
Unknown	
Final Line Busy	38
Final No Answer	525
Language Barrier	25
Total Unknown	588
Total Dispositions	3750

CASRO	11.63%
Upper Bound	16.69%
Refusal Rate	26.61%

6. Midwest 2

Household	
Quota Cell Full	19
Complete	197
Refusal (Eligibility Unknown)	1185
Final Termination in Questionnaire	16
No Eligible Respondent During Interview	228
No Eligible Respondent at Number	4
Total Household	1649
Non-Household	
Disconnect/ Non-working	1222
Not a Private Residence	924
Total Non-Household	2146
Unknown	
Final Line Busy	23
Language Barrier	62
Final No Answer	720
Total Unknown	805
Total Dispositions	4600

CASRO	9.86%
Upper Bound	14.09%
Refusal Rate	26.11%

7. Midwest 3

Household	
Complete	197
Quota Cell Full	281
Refusal (Eligibility Unknown)	1245
Final Termination in Questionnaire	26
No Eligible Respondent During Interview	187
Total Household	1936
Non-Household	
Not a Private Residence	737
Disconnect/ Non-working	1095
Total Non-Household	1832
Unknown	
Don't Know	5
Final Line Busy	35
Language Barrier	27
Final No Answer	610
Total Unknown	677
Total Dispositions	4445

CASRO	8.63%
Upper Bound	13.42%
Refusal Rate	28.59%

8. East 1

Household	
No Eligible Respondent During Interview	269
Final Termination in Questionnaire	57
Refusal (Eligibility Unknown)	1239
Complete	200
Quota Cell Full	97
No Eligible Respondent at Number	1
Total Household	1863
Non-Household	
Not a Private Residence	519
Disconnect/ Non-working	944
Total Non-Household	1463
Unknown	
Don't Know	1
Final No Answer	678
Language Barrier	75
Final Line Busy	64
Total Unknown	818
Total Dispositions	4144

CASRO	8.62%
Upper Bound	13.37%
Refusal Rate	31.27%

9. East 2

Household	
No Eligible Respondent at Number	3
No Eligible Respondent During Interview	286
Final Termination in Questionnaire	46
Refusal (Eligibility Unknown)	1347
Complete	211
Quota Cell Full	123
Total Household	2016
Non-Household	
Disconnect/ Non-working	1326
Not a Private Residence	884
Total Non-Household	2210
Unknown	
Final Line Busy	36
Language Barrier	122
Don't Know	1
Final No Answer	639
Total Unknown	798
Total Dispositions	5024

CASRO	8.80%
Upper Bound	13.15%
Refusal Rate	27.73%

10. Southeast 1

Household	
Complete	195
Final Termination in Questionnaire	45
No Eligible Respondent at Number	5
No Eligible Respondent During Interview	509
Refusal (Eligibility Unknown)	1903
Quota Cell Full	271
Total Household	2928
Non-Household	
Not a Private Residence	1075
Disconnect/ Non-working	1479
Total Non-Household	2554
Unknown	
Final Line Busy	118
Language Barrier	185
Final No Answer	1062
Total Unknown	1365
Total Dispositions	6847

CASRO	5.33%
Upper Bound	9.10%
Refusal Rate	28.45%

11. Southeast 2

Household	
No Eligible Respondent During Interview	303
Quota Cell Full	110
Complete	200
Refusal (Eligibility Unknown)	863
Final Termination in Questionnaire	14
No Eligible Respondent at Number	4
Total Household	1494
Non-Household	
Disconnect/ Non-working	1169
Not a Private Residence	754
Total Non-Household	1923
Unknown	
Language Barrier	85
Final Line Busy	41
Final No Answer	457
Total Unknown	583
Total Dispositions	4000

CASRO	11.44%
Upper Bound	18.57%
Refusal Rate	21.93%

12. South

Household	
Quota Cell Full	27
Refusal (Eligibility Unknown)	1106
Final Termination in Questionnaire	32
Complete	185
No Eligible Respondent During Interview	153
No Eligible Respondent at Number	2
Total Household	1505
Non-Household	
Not a Private Residence	372
Disconnect/ Non-working	208
Total Non-Household	580
Unknown	
Final No Answer	456
Language Barrier	16
Final Line Busy	34
Total Unknown	506
Total Dispositions	2591

CASRO	9.89%
Upper Bound	13.98%
Refusal Rate	43.92%

APPENDIX E
MANAGER'S SURVEY INSTRUMENT

Respondent Name and Position:

Utility Name:

Utility capacity:

Source water:

Source water classification:

☐ groundwater ☐ surface water

☐ blended (% of each)

Utility classification:

☐ public ☐ private

Geographical location:

Population served:

Survey Questionnaire:

The following list of questions is intended to identify your perceptions as a utility representative of how your customers feels about their tap water and how they feel about bottled water and filtration devices. These questions have been structured to compliment the data to be collected in the customer telephone survey. The questions have been designed to elicit quick responses. However, please feel free to elaborate on any of the issues, as you feel appropriate.

1. Which of the following do you believe best describes your average customers satisfaction with the **overall quality** of the tap water they receive in their household. Would you say they were…?

☐ very dissatisfied
☐ dissatisfied
☐ neither dissatisfied nor satisfied
☐ satisfied
☐ very satisfied

2. Which of the following do you believe best describes your average customers satisfaction with the **taste** of the tap water they receive in their household? Would you say that they were…?

- ☐ very dissatisfied
- ☐ dissatisfied
- ☐ neither dissatisfied nor satisfied
- ☐ satisfied
- ☐ very satisfied

3. Which of the following do you believe best describes how **safe** the average customer believes the tap water they receive in their household is to drink? Would you say they believe their water is…?

- ☐ unsafe
- ☐ somewhat unsafe
- ☐ neither unsafe nor safe
- ☐ somewhat safe
- ☐ safe

4. Which of the following do you believe best describes how your customers feel about the **healthiness** of the tap water they receive in their household? Would you say that they believe the water is…?

- ☐ healthy
- ☐ somewhat healthy
- ☐ neither healthy nor unhealthy
- ☐ somewhat unhealthy
- ☐ unhealthy

5. Considering the following water quality features which one do you believe is the most important to your customers? Would you say it was the waters…?

- ☐ Taste
- ☐ Appearance
- ☐ Smell
- ☐ Safety
- ☐ Healthfulness
- ☐ Other ______________________________

6. Do you believe there is a significant difference in how people who drink tap water and people who drink bottled water perceive the taste of their tap water?

What about the smell of their tap water?

7. Do you believe there is a significant difference between how tap water drinkers and bottled water or filtered water drinkers perceive the healthiness of the tap water that they receive in their household?

What about the safety?

8. How sensitive are your average consumers to chlorinous taste and odor in their tap water…?

- ☐ sensitive
- ☐ somewhat sensitive
- ☐ neither sensitive nor insensitive
- ☐ somewhat insensitive
- ☐ insensitive

9. Do your consumers believe that chlorinous taste and odor is a significant problem for their tap water?

10. Do other water quality aspects (*e.g.* hardness, color) influence people's perceptions of chlorinous flavors in their tap water?

11. Do you believe your consumers appreciate/understand the benefits of chlorine in tap water, or is public education required/needed?

12. Have you investigated methods of reducing chlorinous taste and odor in your finished water?
If so, were they feasible solutions?

13. Do you have plans to change/improve public perceptions of the chlorinous taste and odor in their water?

14. Of the following water quality issues, which do you believe is the most common reason for customers to choose to use a water filtration system?

- ☐ Taste
- ☐ Appearance
- ☐ Smell
- ☐ Safety
- ☐ Overall quality
- ☐ Other ______________________________

15. Of the following water quality issues, which do you believe is the most common reason for customers to choose to use bottled water?

- ☐ Taste
- ☐ Appearance
- ☐ Smell
- ☐ Safety
- ☐ Overall quality
- ☐ Other ______________________________

16. Indicate the categories in which you believe that the bottled water consumers in your service area believe bottled water is better than tap water.

- ☐ Taste
- ☐ Smell
- ☐ Safety
- ☐ Healthiness
- ☐ Quality
- ☐ Cost

17. Which of the following best describes your average customer's overall satisfaction with their municipal water provider? Would you say they were…?

- ☐ very dissatisfied
- ☐ dissatisfied
- ☐ neither dissatisfied nor satisfied
- ☐ satisfied
- ☐ very satisfied

18. Which of the following best describes your average customer's satisfaction with the customer service provided by the water utility? Would you say they were…?

- ☐ very dissatisfied
- ☐ dissatisfied
- ☐ neither dissatisfied nor satisfied
- ☐ satisfied
- ☐ very satisfied

19. Which of the following best describes an average customer's level of satisfaction with the resolution of questions or complaints which they may have proposed. Would you say they were…?

- ☐ very dissatisfied
- ☐ dissatisfied
- ☐ neither dissatisfied nor satisfied
- ☐ satisfied
- ☐ very satisfied

20. Do you believe your average customer is aware that they have received a Drinking Water Quality Report from the water utility within the last year?

Is so, do you believe that they found this report useful?

21. Is the report shipped separately or along with a water bill (circle the appropriate answer)?

Separate Along with bill

22. Do you believe that sending out water quality reports is a useful way to educate the public about their drinking water quality?

23. Do you believe there may be a better way to achieve this aim?

If so, what is it?

24. Within the last year, were there any local incidents that seriously threatened the water quality of the water that customers received in their household?

If so, can you described the nature of this incident?

25. Are you aware of any local incidents within the last year that may have caused customers to perceive a threat to the quality of the water they receive in their household, although there was no actual real threat, as in a negative media report or water quality problems in other places?

If so, can you describe this/these event(s)?

26. Are most customer water quality complaints what you would consider legitimate (*e.g.* not due to problems within the home, misunderstandings about water quality, or the result of perceived problems)?

27. Are you concerned about the public's perceptions of tap water?

What aspects of these perceptions concern you most?

28. Does your utility have other PR projects besides CCRs to actively improve the public perception of your tap water or your utility in general?

29. How would you feel about operating / supplying bottled water or POU/POE devices through your utility?

30. Is this something you would seriously consider initiating if it were shown to be legally and logistically feasible?

Thank you for participating in our survey.

As we stated previously, your responses will be kept totally anonymous and only descriptors such as utility size and water source will be used to describe you. If you have any further questions, or are interested in the outcome of the survey, please feel free to give me a call. Thank you again for your time.

APPENDIX F
SUMMARY OF CONSUMER SURVEY DATA

A very large number of comparisons and correlations could be performed on the collected raw data, but size and manageability constraints limit the practical size of the report. In order to facilitate further manipulation of the collected data by interested parties, excel files of the data have been prepared on the following topics:

1. Raw Data
2. Market Data
3. Cross-Correlations By Market

Raw Data

The raw data collected during this project is stored in the excel file "Raw_Data_All_Markets.xls." The "data dictionary" provided below describes the encoding of the raw data files.

Data Dictionary

This dictionary of coding terms accompanies excel file Raw_Data_All_Markets.xls stored on the accompanying CD-ROM.

Value	Label
1.00	West 1 Market
2.00	East 2 Market
3.00	Midwest 1 Market
4.00	Southeast 1 Market
5.00	Southeast 2 Market
6.00	Midwest 3 Market
7.00	Southwest Market

(Continued)

(Continued)

Value	Label
8.00	Midwest 2 Market
9.00	West 2 Market
10.00	East 1 Market
11.00	South Market
12.00	West Coast Market

Market Data

The weighted summaries for each market surveyed are provided on the accompanying CD-ROM. This section describes the data stored in the "Market Data 'Code Letter'.xls" files, also stored on the accompanying CD-ROM. The markets are coded by letter:

Market	Code letter
West 1	A
East 2	B
Midwest 1	D
Southeast 1	E
Southeast 2	F
Midwest 3	G
Southwest	J
Midwest 2	K
West 2	L
East 1	M
South	N
West Coast	O

For instance, the summary data for the market Midwest 1 is stored in file "Market Data D.xls." Data for all markets is stored in file "Market Data for combined markets.xls". This data file is different than the Raw Data discussed in the previous section as all the consumer responses have been decoded from a number to a descriptor (e.g., "1" is equivalent to a "yes" response).

Each data file contains two worksheets, XTABS and FREQS. The "Xtabs" worksheet breaks each market into 2 groups - tap water drinkers and bottled/filtered water drinkers. Each category lists the number of persons responding to each of the options (very satisfied/satisfied/ neither satisfied nor dissatisfied/..., taste/smell/appearance/..., yes/no, etc.). The "FREQS" worksheet looks at the combined group answers. "Frequency" again lists the number of persons responding to each of the options, "Percent" shows the percentage responding to each Frequency category, "Valid percent" shows the percentage responding to each valid Frequency category (answers such as "refused" or "I do not know" are considered invalid and so are not counted), "Cumulative Percent" shows the accumulated percent responding to valid answers.

Survey Response Cross-Correlations By Market

A very large number of cross-correlations were performed on the collected market data. Only the significant correlations were discussed in the text of the document. In order to facilitate further examination and manipulation of the collected data by interested parties, excel files of the correlations are provided on the accompanying CD-ROM.

Answer Key for Cross-Correlations

This section describes the data stored in the Cross Correlations X.xls files, also stored on the accompanying CD-ROM. Correlations were run for a number of the survey questions, listed in the "Variable" column of each data file. The markets are coded by the same code letter discussed previously:

For instance, the cross-correlations data for the market Midwest 1 is stored in file "Cross-correlations D.xls." Correlations between consumer groups (TW, BW, and FW drinkers) and utility types are stored in files "Cross-correlations between consumer groups.xls" and "Cross-correlations between utility types.xls" respectively.

The market data correlations are provided in 7 worksheets:

Age - Test differences between age groups (Q220)
Gender - Test differences between genders (Q230)
Education - Test differences between education groups (Q250)
H[ouse]H[olds] w[ith]_child[ren] - Test differences between HH with/without children (Q255)
QuotaQ - Test differences between primarily TW/BW-FW drinkers (QuotaQ)
BW_drinker - Test differences between BW-FW drinkers (Q095)
TW_drinker - Test differences between TW drinkers (Q035)

The correlations between consumer groups are separated by questions. For example, "BW_drinker" worksheet used Q095 to correlate bottled water consumers (vs. non-bottled water consumers) with satisfaction and demographics. Correlations between utility types are found in a separate file and analyzed based on ownership status, size and source water.

Column A lists the questions in the survey, column B describes the question, column C lists the type of test used, column E identifies the questions tested, column F lists the test statistic result, column G lists the p value (values greater than 0.05 had no significant correlation), the column H lists the ε^2 value and column I describes the type of difference found where correlations were significant.

REFERENCES

Auslander, B.A., and P.H. Langlois. 1993. Toronto Tap Water: Perception of its Quality and use of Alternatives. *Can. J. Public Health*, 84:99–102.

AWWA (American Water Works Association). 2001. www.awwa.org/pressroom/statswp5.htm. INTERNET.

AwwaRF (Awwa Research Foundation). 2003. Public Perception of Tap Water Chlorinous Flavor. *Work in Progress.*

Bachmann, T. 2000. Pouring it on. *IBWA Annual Convention and Trade Show.*

Bottled Water Web. 2000. www.bottledwaterweb.com/statistics.html. INTERNET.

Burlingame, G.A., J.J. Muldowney, and R.E. Maddrey. 1992. Cucumber Flavor in Philadelphia's Drinking Water. *Jour. AWWA*, 84(4):92–97.

Burlingame, G.A., R.M. Dann, and G.L. Brock. 1986. A Case Study of Geosmin in Philadelphia's Water. *Jour. AWWA*, 78(3):56–61.

CDC. 2002. Surveillance for Waterborne-Disease Outbreaks -- United States. 1993-1994. www.cdc.gov; INTERNET.

ChillCan. 2001. About the Can. www.chillcan.com; INTERNET.

Cochran, W.Q. 1977. Sampling Techniques. 3rd Edition. New York, N.Y.: Wiley.

Doulton Water Filters. 2001. Point-Of-Use Technologies and Applications. www.doulton.ca.

Dysard, J.A. II. 1999. How Competition is Changing the Face of the Public Water Resources Industry Trends in Privatization, Managed Competition, and Other Alternative Delivery Systems. Chicago, Ill. In *Proceedings of AWWA Water Quality Technology Conference.* Denver, Colo.: AWWA.

EPA (Environmental Protection Agency). 2001a. *Envirofacts Warehouse.* www.epa.gov; INTERNET.

EPA (Environmental Protection Agency). 2001b. Factoids: Drinking Water and Ground Water Statistics for 2000. www.epa.gov/safewater; INTERNET.

Ferrick, T. 2000. A Different Danger - Lead in the Water - Stalks Philadelphia Schools. B1, B6. Oct. 1, 2000.

Gallagher, M.W. 2000. Professional Selling Skills. *International Bottled Water Association Annual Convention*. Seattle, Wash.: IBWA.

Garrett Specialties. 2001. Bottles. Available at www.garrettspecialties; INTERNET.

Gevirtz, L. 1999. Philadelphia Water Bad for Elderly's Health. www.reuters.com; INTERNET.

Haaga, J. Kent M.M., Mather, M., Pollard, K.M. June 2001. First Glimpses from the 2000 U.S. Census. *Population Bulletin*, 56(2).

Hall, E. 2000. A Brief History of Drinking Water. *Annual AWWA Newsletter* (VA. Section)

Hall, S. and Miller, E.G. 2001. Kansas City's Bottled Water Experience. *Opflow* 26(1), 4.

Hellard, M.E., M.I Sinclair, A.B Forbes, and C.K. Fairley. 2001. A Randomized, Blinded, Controlled Trial Investigating the Gastrointestinal Health Effects of Drinking Water Quality. *Environmental Health Perspectives,* 109(8):773-778.

Henry, G.T. 1990. Practical Sampling. Newbury Park, Calif.: Sage.

Holleran, J. September 2000. Bottled Water Comes of Age. *Beverage Industry,* 22–27.

Hurd, R. E. 1993. *Consumer Attitude Survey on Water Quality Issues*. Denver, Colo.: AwwaRF and AWWA

Jaffe, M. 1995a. Parasite Discovered in City Drinking Water. *The Philadelphia Inquirer*, Al, A19.

Jaffe, M. 1995b. Regional Drought Warning Looms. *The Philadelphia Inquirer*, B1, B3.

Krasner, S.W., and S.E. Barrett. 1984. Aroma and Flavor Characteristics of Free Chlorine and Chloramines. *Proceedings of AWWA Water Quality Technology Conference*. Denver, Colo.: AWWA.

Lyonnaise des Eaux. Juillet 1998. *Perception des Goûts et Odeurs de l'eau*. Rapport d'étude. Centre Operationnel des Yvelines: SOFRES.

Mackey, E.D., J. Davis, and G., Crozes. 2001. Customer Attitudes and Perceptions of Point-of-Use Applications and Bottled Water Use. *Proceedings of the AWWA Annual Conference*. Denver, Co.: AWWA.

McKinney, M.P. 2000. Contaminated Water Found at BHC Company. The News-Time (online), June 30. www.newstimes.com; INTERNET.

NewsMax. 2001. Worst Drought in History Leaves Florida High and Dry. www.newsmax.com. INTERNET. March 1, 2001.

Oanh Ha, K. 2001. Bottled Water Standards – Just How Safe is Filtered Water. *San Jose Mercury News*.

Olson, E.D. 1999. Bottled Water: Pure Drink or Pure Hype? www.nrdc.org; INTERNET.

PopStraw. 2001. The products. www.popstraw.com; INTERNET.

Seattle Times. 2000. Big Market Seen for Greenland's Top Product: Water. www.seattletimes.com, INTERNET.

Suffet, I.H. 2001. Bottled Water. Southern California Environmental Report Card 2001. UCLA Institute of the Environment. 32-41.

Suydam, J.A. 1999. City's Murky Tap Water Too Much for Most on Council. Corpus Christi Online. www.caller.com; INTERNET.

Sydney Water. 2001. Community Views on Drinking Water Quality. Research Report.

Universal Studios. 2000. www.screenit.com/movies/2000/erin_brokovich.html; INTERNET.

Viviano, J. 1998. Chemical Spill Could Threaten Water Supply. *The New Haven Register*, A1, A4. July 6.

Water Quality Association. 1999. 1999 National Consumer Water Quality Survey. www.wqa.org; INTERNET.

Water Quality Association. 2001. 2001 National Consumer Water Quality Survey. Available by request at www.wqa.org; INTERNET.

WaterStats. 1996. Water Statistics Database for U.S. Utilities. AWWA and AwwaRF.

WaterTechOnline. 2000. Survey: US Drinking More Bottled Water. www.watertechonline.com; INTERNET.00000

WaterTechOnline. 2001a. Bottled Water Sales Bubbling the UK. www.watertechonline.com; INTERNET.

WaterTechOnline. 2001b. WQA Chief Meets with Congressional Leaders. www.watertechonline.com; INTERNET.

Yantis, J. 2001. East Valley's Supply Safe - It Just Tastes Terrible. *East Valley Tribune*, A1, A8. March 18, 2001.

ABBREVIATIONS AND VARIABLES

ANSI	American National Standards Institute
ASR	aquifer storage and recovery
AWWA	American Water Works Association
AwwaRF	Awwa Research Foundation
BRFSS	behavioral risk factor surveillance system
BW	bottled water
°C	degrees Celsius
$CaCO_3$	Calcium carbonate
CASRO	Council of American Survey Research Organizations
CATI	computer-aided telephone interviewing
CCR	consumer confidence reports
CDC	Centers for Disease Control
DBP	disinfection by-product
EPA	Environmental Protection Agency
°F	degrees Fahrenheit
FAQ	frequently asked questions
FDA	Food and Drug Administration
FW	filtered water
GIS	graphical information system
GW	ground water
HAAs	haloacetic acids
IBWA	International Bottled Water Association
L	liter
%	percent
MCL	maximum contaminant level
µg/L	microgram per liter
mg/L	milligrams per liter
mgd	million gallons per day
MRA	Marketing Research Association
MTBE	methyl ter-butyl ether

NEETF	National Environmental Education & Training Foundation
NRDC	National Resources Defense Council
NSF	National Sanitary Foundation
NTU	nephelometric turbidity unit
POU/POE	point-of-use/point-of-entry
PPM	part per million
PR	public relations
RDD	random digit dialing
RO	reverse osmosis
SPSS	Statistical Product and Service Solution
SW	Surface water
T&O	taste and odor
TDS	total dissolved solids
THMs	trihalomethanes
TOC	total organic carbon
TTHM	total trihalomethanes
TW	tap water
TWA	tap water alternatives
U.S.	United States
USEPA	United States Environmental Protection Agency
UV	ultraviolet
WQA	Water Quality Association
WTPs	water treatment plants

IWA Publishing's authorised EU representative for General Product Safety Regulations is Diane D'Arras, 15 rue Duret, 75116 Paris, France, e-mail: safety@iwap.co.uk.

Printed and bound by CPI Group (UK) Ltd, Croydon, CR0 4YY
17/09/2026
02218808-0019